Textile Techniques in METAL

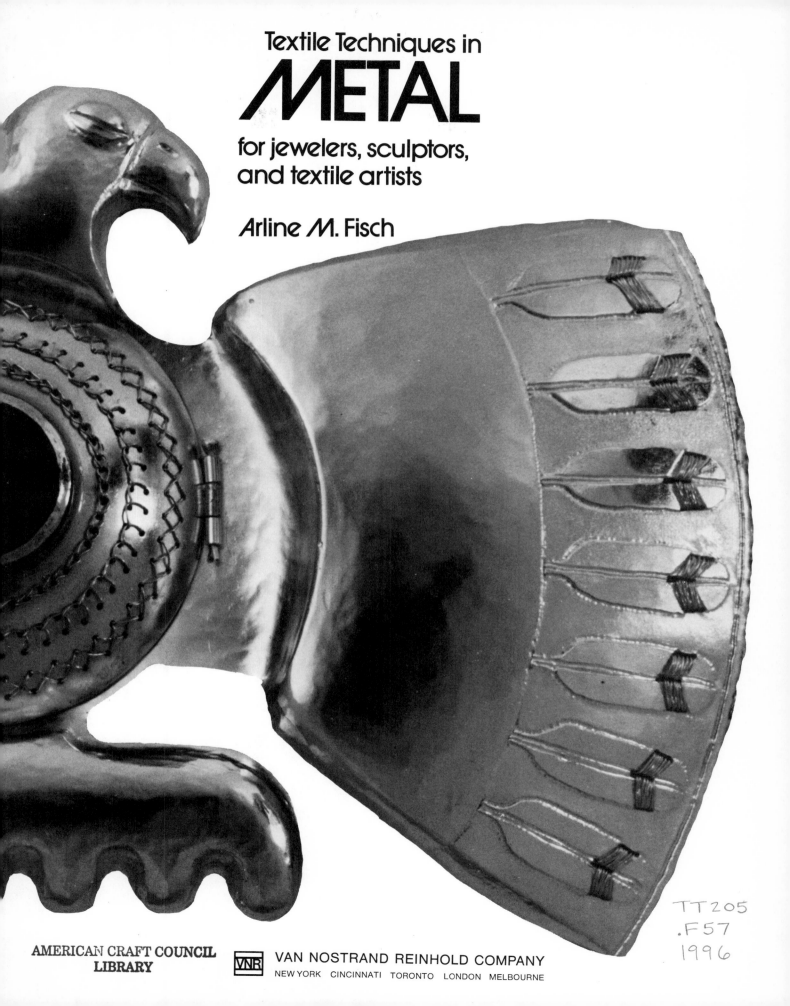

Textile Techniques in
METAL
for jewelers, sculptors,
and textile artists

Arline M. Fisch

VNR VAN NOSTRAND REINHOLD COMPANY
NEW YORK CINCINNATI TORONTO LONDON MELBOURNE

To my family; their love, encouragement, and moral support have made all things possible

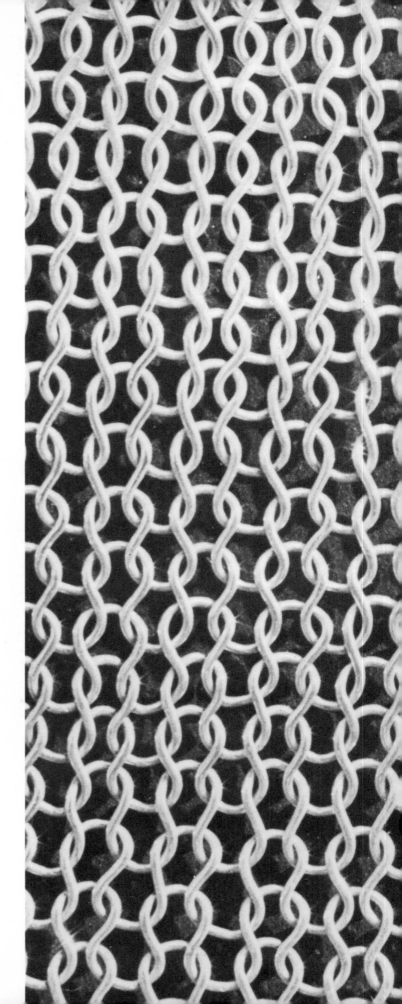

Page 1

Detail of *Two-headed Bird* pendant by Randy Long.

Pages 2 and 3

Two-headed Bird, chased silver pendant by Randy Long. Chain stitch, buttonhole stitch, cross-stitch, and satin stitch in 18-karat gold are worked through holes drilled in wings and hinged container door.

Van Nostrand Reinhold Company Regional Offices:
New York Cincinnati Chicago Millbrae Dallas
Van Nostrand Reinhold Company International Offices:
London Toronto Melbourne

All samples and examples of finished pieces are my own work unless otherwise indicated.

Color photography by Michael Chesser, Los Angeles.
Designed by Loudan Enterprises

Published by Van Nostrand Reinhold Company
A Division of Litton Educational Publishing, Inc.
450 West 33rd Street, New York, N.Y. 10001

16 15 14 13 12 11 10 9 8 7 6 5 4 3 2 1

Library of Congress Cataloging in Publication Data
Fisch, Arline.
   Textile techniques in metal for jewelers, sculptors,
and textile artists.
   Bibliography: p.
   Includes index.
   1. Metal-work. 2. Textile crafts. 3. Metal cloth.
I. Title.
TT205.F57   746'.04'53   74-15249

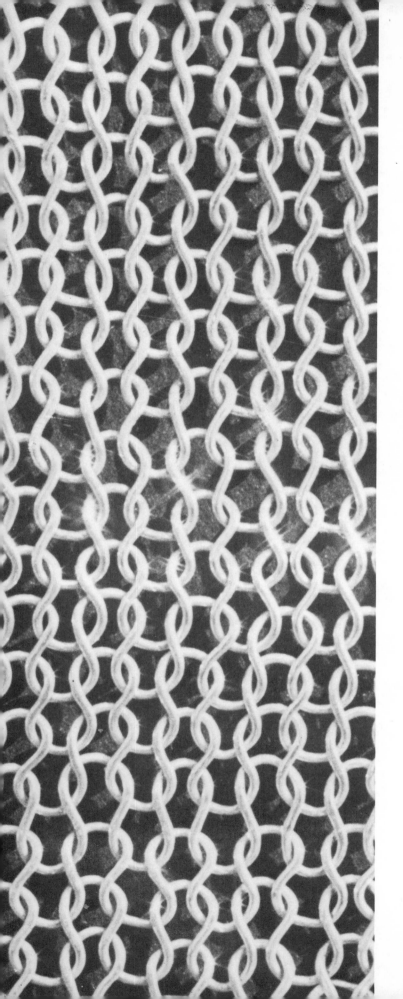

Acknowledgments

I wish to acknowledge the assistance of the curatorial staff of various museums whose help in locating historical material was invaluable. These include: The San Diego Museum of Man, Cooper-Hewitt Museum of Design and The National Museum of Natural History of The Smithsonian Institution, The Lòs Angeles County Museum of Art—Textile Collection, The Museum of Primitive Art, The Field Museum of Natural History, The Cleveland Museum of Art, and The National Museum of Finland.

I wish to express very special thanks to my many colleagues who so generously shared their work and their technical experience with me; to my students at San Diego State University who contributed their work and their enthusiasm to this project; to the many craftsmen whose ideas and suggestions on the subject stimulated my thinking; to the World Crafts Council which introduced me to craftsmen in many countries whose work and interest in the subject gave me the courage to undertake this book; and to my friends who so patiently shared both my trials and my tribulations throughout this endeavor.

Detail of necklace in flat-knitted silver mesh by Richard Anderson.

Contents

Neckpiece by Marci Zelmanoff in fine silver and yellow linen. Techniques include weaving, wrapping, and macramé. The knotting of the wire for the macramé shows that it is possible to make complex knotted structures in wire, although it stresses the metal beyond its natural capabilities.

". . . People who start a handicraft go unprejudiced to the task, manage as best they can with the aids available, and then make arrangements for warding off difficulties. Tradition is created little by little."

(*Olddanske Tekstiler* by Margrethe Hald)

Three Fishermen, chased silver pendant by Richard Anderson, with dimensional net of knitted silver mesh.

Tubular knitted structures in silver and copper by Richard Anderson.

Introduction

Sitting on the deck of a Turkish freighter in the harbor at Marseilles with the prospect of four days of total idleness before debarking in Piraeus was supposed to have a relaxing effect on my tired brain and body. Instead, it was terrifyingly boring. I don't take to idleness well, and the thought of having nothing to do made me feel very restless. Luckily the ship's captain was an accommodating man quite used, I suppose, to the curious requests of his multinational passengers. He politely suppressed any surprise he felt at my request, and within the hour presented me with a huge coil of small-gauge copper wire, courtesy of the radio operator. I spent the trip playing with ways to structure wire without benefit of tools or equipment, and the result has been a continuing fascination with the transposition of textile processes into metal.

The idea that metal can be structured in the same manner as textiles is certainly not original. It has existed for centuries in many different places and cultures. My first encounter with it came when looking at pre-Columbian textiles and goldwork. The first thing I noticed was that metal objects could be sewn to garments permanently instead of being totally separate (or temporarily attached) ornaments meant to be worn on garments. I later found many other examples of this permanent relationship between fabric and ornament in cultures as varied as the Viking and Scythian. My second observation was that cast-gold ornaments often resembled textiles in their surface patterns, indicating that models for the cast might well have been made of fibers and yarn impregnated with wax and then braided or twisted to form strong textile patterns.

Gradually, as I explored various ways of incorporating metal and fiber elements—first in the mere sewing of metal ornaments to fabric forms, and then to an integration of metal into woven yarn structures—it occurred to me that neither the fabric nor the yarn structures were necessary, and that the same textural and structural effects could be accomplished directly in metal, particularly wire. My shipboard explorations had shown me that I could weave, twine, and braid with wire to produce self-supporting structures without benefit of any metalworking process.

Next, I saw a small fragment of woven gold in a private collection of pre-Columbian goldwork in Peru. It consisted of small ribbons of thin sheet gold interlaced in a plain-weave structure, and even in its fragmented state its light-reflective qualities were absolutely beautiful. I have been involved ever since, experimenting with all types of textile structures and techniques.

For a craftsman with enough patience and resourcefulness, I believe *all* textile methods would be possible to execute in metal. Some techniques, however, force the material beyond its natural workability, causing problems in the process that I do not enjoy. For that reason, I have deliberately excluded processes like knotting and macramé, which require movements in opposition to the natural behavior of metal. They can be used, of course, if the material is handled very gently and carefully and if it is sufficiently fine in size, but the danger of kinks and breaks during the intricate interlacing and necessary tightening is constant and discouraging. Embroidery also pushes metal into uncomfortable positions and circumstances. Nevertheless, I have included photographs of work by other craftsmen who have used difficult techniques such as these very successfully.

The single-element structures of crochet and knitting are very simple to execute in wire and have the advantage of not requiring any frame or large implements. The fact that they can be constructed directly off a spool of wire makes them both portable and easy to handle. Structures which involve one set of elements, such as braiding, sprang, and bobbin lace, are more complex because they require both a frame or a base on which to work as well as multiple ends of wire or sheet metal strips which must be kept in order. The interlacing structures like weaving, which are two-element techniques, also require equipment, sometimes elaborate, to support the work while it is in process. Basketry, on the other hand, uses two sets of non-supported elements to form its structure.

In describing the various structures, I have referred as much as possible to Irene Emery's *The Primary Structures of Fabrics* for classification and have made use of her terminology in outlining the basic processes. I have not attempted to give detailed directions for any of the techniques mentioned—only as much as might be necessary for a craftsman to begin some experimental samples in wire or sheet metal. There are so many books and instructional manuals available now on each process that to repeat them here in abbreviated form seems redundant. Instead, the reader can refer to the Bibliography at the end of the book, which has been organized by specific techniques.

The finished textile structures have multiple possibilities dependent upon the intention and the capabilities of the maker. They can be executed in any scale, from a ring for the finger to a relief for a wall, or in three-dimensional form from the most delicate beads to free-standing sculpture. The structure can be complete in itself or it can be combined with rigid elements to produce contrast in texture and form. It can be created by textile techniques without any metal technology or constructed by traditional metalworking with the textile-work serving as an ornamental device. There is no one right way; there is only the artist choosing which direction he will take.

That many artists do choose to combine metal and textile technologies is evidenced by the number who are now involved with the processes. An American sculptor made knitted wire hanging structures twenty years ago; a Norwegian designer makes knitted silver-wire bracelets now; a British student spool knitted a necklace ten years ago; a contemporary American student uses cloisonné wire to produce four-strand braided beads. As so often happens, artists in different times and different places stumble onto the same idea seemingly without reference to each other or to historical antecedents. Using textile techniques in metal does not constitute a movement or a style, only a shared interest in an exciting concept that is somehow "in the air" and has inexhaustible possibilities.

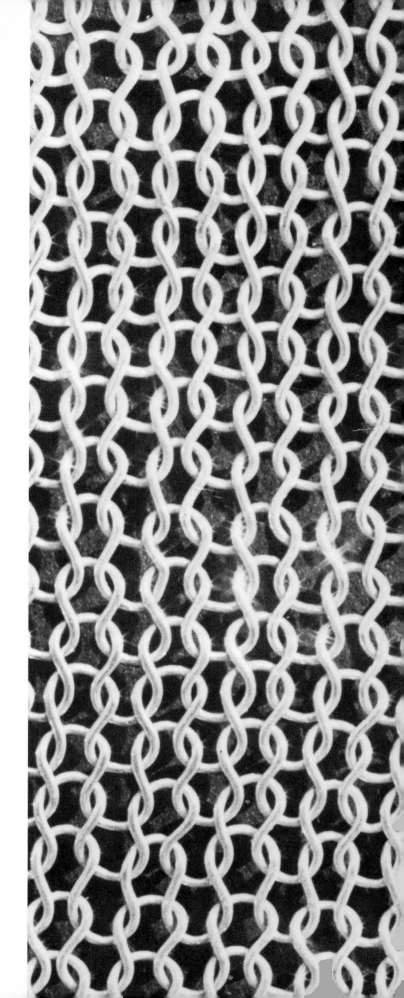

Detail of necklace in flat-knitted silver mesh by Richard Anderson.

1/Historical Precedents

". . . And made the holy garments for Aaron, as the Lord commanded Moses . . . and they did beat the gold into thin plates and cut it into wires, to work it in the blue, and in the purple, and in the scarlet, and in the fine linen, with cunning work . . . Chains of pure gold, twisted like cords, were made for the breast piece . . . " (Ex. 39:1-15)

Examples of garments and artifacts combining metal and textile technologies can be found in cultures as disparate in time and location as ancient Greece, pre-Columbian Peru, medieval Europe, seventeenth-century Japan, Sung and Mandarin China, primitive Africa, and twentieth-century industrial societies. They do not evidence any particular chronology nor do they represent any cultural, stylistic, or technological development. Rather, they seem to occur randomly whenever sophisticated technologies in both metalwork and textiles exist simultaneously.

There seems to be a large quantity of overlooked and neglected work of this type, but it is not easy to locate. Except for the extensive use of metal-covered threads employed in the making of laces, brocades, and elaborate embroideries, most textile techniques in metal are usually isolated techniques found in unrelated cultures. The actual substitution of metal (in wire or sheet form) for yarn or fabric does not appear often in the making of clothing, and examples are not usually found in textile collections, although sometimes one sees metal textiles used as ornamental devices fastened to garments. More often, the processes are used to ornament larger objects or to produce entire small objects such as baskets.

Collecting information and locating examples has been a fascinating and frustrating experience for me, and the results are certainly not an exhaustive survey of the subject. That task I leave to someone more qualified than I to engage in thorough historical research. Instead, I am content to share those examples which have crossed my path in one way or another, and whose discovery has stimulated my thinking, provoked my curiosity, and fueled my enthusiasm for contemporary application of a dual technology. I have relied completely on my own observations and that of friends, acquaintances, and willing museum personnel to uncover areas for consideration as well as specific examples.

Some materials fall into a nonstructural category in which metal elements are sewn or otherwise fastened directly to fabric structures to produce a richly decorative effect. A pre-Columbian poncho from the Ica culture of the south coast of Peru (circa 1000 to 1470 A.D.) is a cotton tapestry of elaborate pattern using motifs of trees and animals (1-1). The lower section is completely covered with small squares of thin sheet gold sewn to the fabric with no space left between the plaques. Each plaque is embossed around the edges and has holes drilled in each corner, allowing them to be sewn to the fabric individually. There are numerous examples of this application of gold to clothing in pre-Columbian collections, indicating that it was widely used, probably for ceremonial occasions. A priest or king so arrayed, standing in the brilliant sunshine of Peru, must literally have dazzled the crowd before him.

Collections of ancient Scythian metalwork are replete with small geometric motifs and figures of animals, pressed in thin sheet gold, that have loops or holes for sewing onto long-since deteriorated fabrics. Preserved painted images on pottery and repoussé images in metal bear witness to the fact that Scythian clothing was indeed embellished with borders of such figures arranged in close sequence.

1-1. A pre-Columbian poncho from the Ica culture of the south coast of Peru (1000–1470 A.D.), woven of cotton with gold plaques sewn on one side, 20 x 46½ inches in folded dimension. Courtesy of The Los Angeles County Museum of Art, Costume Council Funds.

1-2. Detail of border ornament from a Finnish apron dated 1200 A.D. Tubes made of spirals of thin bronze wire are sewn on the cloth in a rosette pattern. Courtesy of The National Museum of Finland, Helsinki.

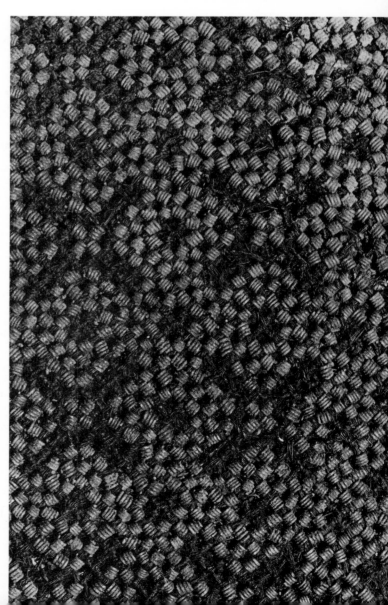

A less spectacular but equally fascinating variation is a woman's costume from Finland dating from about 1100 A.D. A woven wool mantle has a border on all four sides which consists of small tubes of metal threaded with yarn and worked into a multi-strand braid which is about one-and-a-half inches in width (1-2). The tubes are probably brass, or some other alloy of copper, and are no longer polished if, in fact, they ever were meant to be. They provide a structural stiffness and weight to the garment, as well as some decorative quality where the same tubular elements are sewn to the dress in medallion-like motifs.

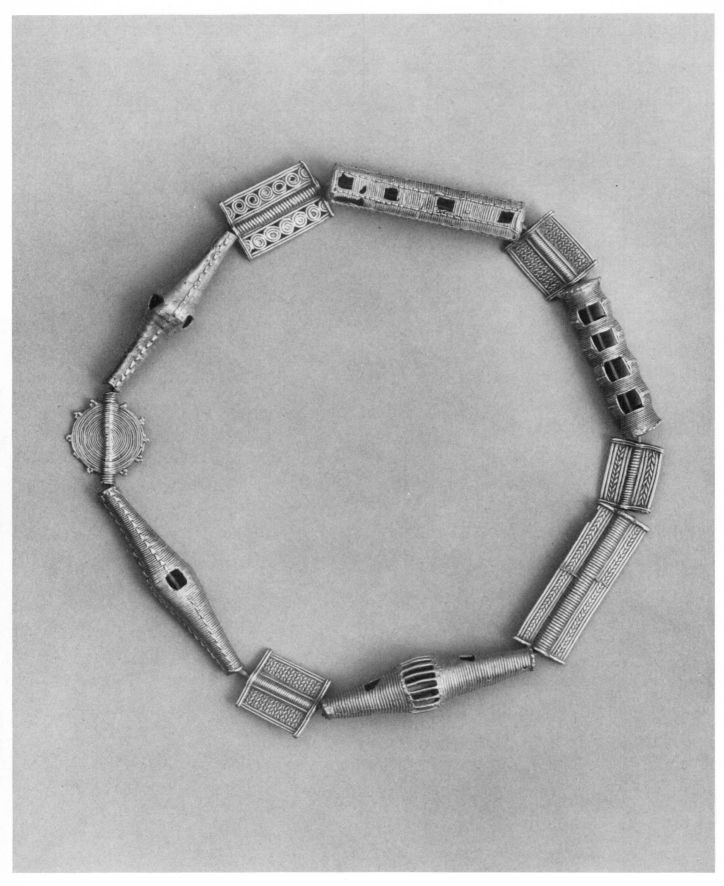

1-3. Necklace of cast-gold beads from the Ashanti region of Ghana, 15¼ inches long. Courtesy of The Museum of Primitive Art, New York.

Rather than structuring metal directly in a textile process, textile patterns and constructions may be cast in metal. This technique has been used in many places but never with more effective results than in two widely different areas, pre-conquest Colombia and Ghana in Africa, the home of the Ashanti. In both cultures, fibers or fibrous materials (probably impregnated with wax) have been twisted and braided to produce highly textured pieces which are then cast into metal by the lost-wax process. An Ashanti necklace of gold beads (1-3) shows an assortment of textile surfaces while a cast-copper pipe of Nigerian origin (1-4) uses a variety of techniques within a single cast structure. Cast-gold ornaments, such as earrings, frequently imitative of wire-filigree patterns, are found in abundance in Colombia while a similar technique is used to fashion small cast animals in both gold and copper in Colombia and Africa (1-5).

1-4. Cast-copper pipe from Nigeria. Courtesy of The San Diego Museum of Man.

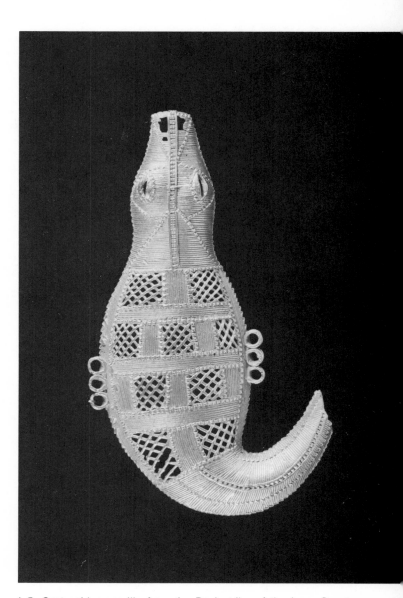

1-5. Cast-gold crocodile from the Baule tribe of the Ivory Coast, Africa, about 4½ x 2 inches. Courtesy of The Cleveland Museum of Art, John L. Severance Fund.

European cultures frequently created textiles using precious metals in rich embroideries and elaborate laces made during the Renaissance and into the eighteenth century. In most of this work the thread was created from metal wire milled down to incredible fineness and then painstakingly wound around a core of yarn, usually silk or linen. It was used to fashion elaborate passementerie borders of silver and gold on Elizabethan handkerchiefs; it was woven into tapestries and embroidered onto the surface of ecclesiastical vestments and royal garments; it was used extensively in making bobbin lace (1-6) of very intricate patterns and great delicacy (see Chapter 7 for modern adaptations). Flat narrow bands of pure metal, again milled to a very thin gauge, were often incorporated into these lace structures. These bands, usually woven into the structure of the metallic thread warp, provided a larger area of metal surface and therefore a greater light-reflective quality (1-7). Three-dimensional effects were occasionally created when warp and weft both were metal bands. Such weaving is elevated from the background surface, much like a small gemstone.

Wire was used to create areas within generally non-metallic lace structures to add stiffness and value. At times purely decorative metal elements like coils are sewn to the surface to add richness and glitter. I have not been able to locate a single example of lace made totally in metal although it seems probable that such a variation must have been made at some time or another. Perhaps it was melted down or unraveled along with so many other metallic-thread laces which were destroyed at a later date for the value of the metal. This practice, called parfilage or drizzling, was so common that a special set of implements was used for removing precious metals from thread—a Victorian method of economic recycling.

1-6. Bobbin-lace ornament, Italian, late sixteenth-century. The lace is made of linen thread, but the spaces are ornamented with wire using the same structure, and the surface is decorated with small metal coils. Courtesy of Cooper-Hewitt Museum of Design, Smithsonian Institution, New York; bequest of Richard C. Greenleaf, in memory of his mother Adeline Emma Greenleaf (#1926-50-198).

1-7. Detail of gold lace, called *fleco de oro,* from Tehuantepec, Mexico, in the Spanish tradition. This type of ornament is still used on traditional garments worn by young girls when they marry. The upper band is woven with thin strips of gold over a yarn warp, while the more complex patterns are made with gold-covered yarn. Courtesy of The San Diego Museum of Man.

A rare and somewhat mysterious piece of so-called "metal lace" comes close to being an entirely woven metal structure (1-8a). It consists of narrow bands woven with a weft of flat metal strips, approximately one millimeter in width, which alternate regularly with a second weft of finer round wire. The warp is cotton spaced in a regular sequence of wide (four to five millimeters) and narrow (one millimeter) thread groupings. The effect is of a totally weft-faced structure in which the flat strips of metal float over the wide warp areas and are locked into place by the inconspicuous wire weft. The reverse side (1-8b) is warp faced with fine wire floats. Little is known of the origin of this technique except that it is undoubtedly European but intended for the American Indian trade. Pieces have been found decorating the shoulder, cuffs, and front opening of a wool shirt unearthed from an Indian grave in South Dakota. The fragment I have described is copper, but there is some evidence that the same technique was used with silver and gilt, perhaps as cheap imitations of military ornamentation. Such woven lace was apparently widely distributed among Indians of the American West during the eighteenth and early nineteenth centuries since it is listed as presents and mentioned in contemporary descriptions of garments worn at Indian ceremonies. Unfortunately, very few actual pieces have been recovered thus far, probably because of the fragile nature of the lace and the fact that most of it was made in copper or copper gilt, which deteriorates rapidly.

1-8a. A fragment of metal "lace" found in an Indian burial in South Dakota (right side). The band is a weft-faced construction of flat strips of copper alternating with thin copper wires woven on a cotton warp.

1-8b. The reverse side clearly indicates the spacing of the cotton warp. Courtesy of The National Museum of Natural History, Smithsonian Institution, Washington, D.C.

Metal directly fashioned in textile construction can be found in a vast range of time, place, and culture, from a small fragment of pre-Columbian origin, which is fabricated of woven strips of gold warp and weft, to a piece of Mycenaean jewelry, presumably a collar, constructed in a basketry technique but using gold wire instead of reed and straw (1-9). Eight rows of the wire are arranged in a continuous spiral, and sturdy wire posts are laced into a vertical position periodically in order to maintain the spacing between rows. Between the verticals, a diamond pattern is formed by the lacing of thin ribbons of gold around each of the horizontal elements in succession, in the same manner as straw would be laced around reed coils to fashion a basket. These diamond motifs are repeated a number of times around the circumference of the collar to produce a continuous pattern which is both decorative and structural.

Oriental metalwork of both Chinese and Japanese origin makes use of textile processes in quite different kinds of objects, as shown in two beautiful gold ornaments of the Sung Dynasty (960 through 1279 A.D.) from two separate crowns. One, a tiny representation of a chrysanthemum blossom with two layers of petals (1-10), is a mere one-and-three-quarter inches in diameter, yet each of the petals on the outer layer is formed of woven gold wire fastened to a rigid outer rim and slightly domed to give a dimensional effect. The other, a *fêng*, or phoenix, of about five inches in length, is constructed totally in the round (1-11). The body, neck, and head of the bird are all fashioned of a fine gold-wire mesh, which is probably twined although the technique is uncertain. An elaborate headdress of the Mandarin period (circa 1850 A.D.) uses basketry techniques on a much larger scale (1-12). The entire crown is formed over a heavy wire frame, which is interlaced with braided bands of silk, to form the main body of the structure. The upper edge is strengthened with a decorative band of open braiding executed in wire completely wrapped with silk thread. The ornaments are made of thin sheet metal covered with brilliant blue kingfisher feathers and multiple gemstones. (Contemporary adaptations of braiding in metal are shown in Chapter 6 and of metal basketry in Chapter 8.)

1-9. A gold collar, perhaps funerary in nature, is a spiral of wire structured and ornamented with lacings of gold in diamond patterns. It is one of two such pieces in the Mycenean collection of The Archaeological Museum, Athens. Courtesy of TAP Service, Athens.

1-10. A gold crown ornament of the Sung Dynasty in the form of a chrysanthemum blossom, about 1¾ inches in diameter. Courtesy of The Metropolitan Museum of Art, New York; Fletcher Fund, 1926.

1-11. A gold crown ornament in the form of a phoenix, Sung Dynasty, about 4½ inches long. Courtesy of The Metropolitan Museum of Art, New York; Rogers Fund, 1929.

1-12. A courtier's headdress from Mandarin China, circa 1850, with a braided wire band around the top of a wire-and-braided fabric structure, 10¼ inches wide x 7½ inches high. Courtesy of The San Diego Museum of Man.

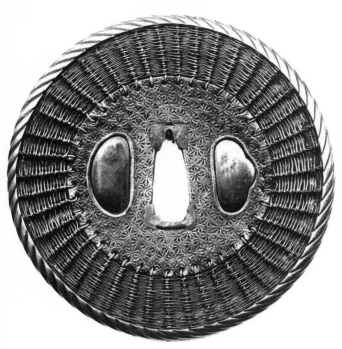

1-13. A Japanese *tsuba*, or sword guard, of the Shingen Mukade style with a woven rim of brass wire. Courtesy of The Field Museum of Natural History, Chicago.

Oriental weapons as well as headdresses make use of textile techniques in both a decorative and structural manner. A Japanese *tsuba*, or sword guard, of the Shingen Mukade style has a brass center and a rim of metal wire that is structurally formed by a radial weaving pattern which is then mounted on an iron disc for greater strength and practicality (1-13). Frequently *tsubas* merely have textile impressions on their surfaces, made either by inlays of different color metals or by textures chased directly on the surface, but it is interesting to see that it did not escape the notice of sword makers that actual textile techniques could be used directly in metal. Other metalworkers occasionally created objects in woven wire, such as a 17th-century Japanese *netsuke* in the shape of a covered container, which has a woven geometric pattern of flat copper wire in black and brown. Apparently the weaving was done in large sections, which were then cut to form the body of the piece and fastened to a strong metal rim.

Other cultures also make some use of textile technology in ornamenting the handles of swords, spears, bows, and arrows. Such decoration on primitive weapons may well be an outgrowth of the earlier practice of binding a metal tip to a wooden shaft, an actual textile process involving the wrapping of wire, cord, or leather thong at the joint and for some distance down the shaft. In two ceremonial weapons from the Shire River District of Africa the wrapping has become highly decorative and covers the shaft completely. The bow (1-14) is densely covered with wire that is woven, wrapped, twisted, and braided in a continuous series of changing decorative structures; the spear (1-15) is totally covered with wire fabrications of an even more complex and decorative nature, with three distinct patterns used in alternating sequence for the entire length of the four-foot shaft. The other spear shown here (1-16), of unknown origin, makes a more structural use of a textile technique. Three heavy forged bands of wire split from a single rod are twisted, braided, and reunited to be forged into the blade.

1-14. A ceremonial bow from the Shire River District of Africa made of wood totally covered with woven, wrapped, twisted, and braided wire. Courtesy of The San Diego Museum of Man, Jessop Collection.

1-15. A ceremonial spear from the Shire River District of Africa covered with woven and braided wire in a variety of decorative patterns. Courtesy of The San Diego Museum of Man, Jessop Collection.

1-16. A large spear of unknown origin in which three strands of metal are twisted and braided to form the shaft. Courtesy of The San Diego Museum of Man, Jessop Collection.

1-17. The handle of a *kris,* or short sword, from the Moro culture of the island of Mindinao, the Philippines. The ornamental bands are of woven silver wire. Courtesy of The San Diego Museum of Man.

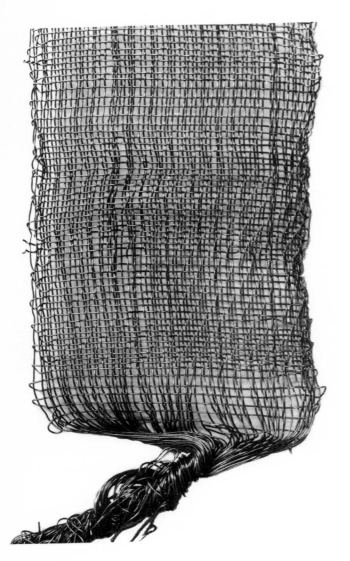

A quite different weapon, a *kris,* or short sword, from the Moro culture of the Philippines, uses a metal version of braided lace bands to ornament the handle (1-17). The decorative bands are actual textile fabrications made directly in silver wire and applied to the silverwork of the body of the sword, much as similar lace bands would be applied to clothing.

Traditional cultures, particularly from the Near East and Middle East, often use textile techniques to fashion metal chains, belts, hats, and wedding regalia. Since these accessories form part of their traditional dress and ceremonies, and their method of manufacture is dictated by tradition, they are hard to date. The forms have not changed even to the present. The unfinished samples of Yemenite metal textiles were gathered and sent to me by an Israeli jeweler who often hires Yemenite metalsmiths in her workshop. They are pieces of recent manufacture, mostly in brass or low alloys of silver, but have evidently been made in the same way for many generations, since the Yemenites use loom-woven and knitted structures in metal to make bracelets, belts, and other ornaments related to their traditional costume.

It is obvious, even though I have no documentation, that the woven strip was made on some kind of loom in a standard twill weave producing a diagonal pattern (1-18). The flat-knitted structure (1-19) must have been done on a frame rather than on straight needles. Each row is wound around a stiff braid on either side, resulting in a structure which is flexible in its length but rigid in its width.

Knitted tubes may be fashioned in many sizes of wire and in varying degrees of fineness (see Chapter 4) to produce a mesh fabric which is probably meant to be flattened by hand first and then rolled through a mill for flexible belt lengths. A belt of Lebanese or Turkish origin illustrates the effect of milling a knitted, interlocked, or woven structure by squeezing and pressing it between steel rollers. This not only flattens the structure but gives a flexibility which creates the appearance of fabric. This process is used extensively in making belts despite the fact that milling weakens the metal where one element crosses another, causing fatigue and breakage with constant use.

1-18. A sample strip of loom-woven wire from Yemen in which a diagonal twill pattern is discernible. All Yemenite samples courtesy of Mimi Rabinowitch of Israel.

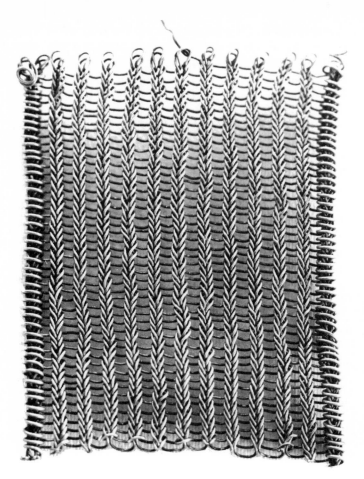

1-19. A flat frame-knitted structure from Yemen.

1-20b. Detail of the belt.

1-20a. A Turkish belt made in an interlocking structure which is then rolled flat to give the appearance of a fabric. Similar belts are found throughout Turkey and Lebanon. Courtesy of Dr. Fay Frick, San Diego.

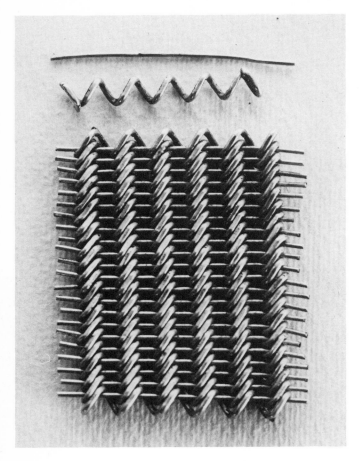

1-21a. A sample of an interlocking structure from Yemen with one row removed to illustrate the method of construction.

1-21b. The finished brooch.

Related techniques which are not continuous textile processes but are made in a non-continuous manner still retain the flexibility of fabric. The small brooch shown here in a finished and unfinished state (1-21a) uses interlocking zigzag coils held in place by rigid weft wires riveted at the end. This technique is similar to that used in the Lebanese or Turkish belt.

A powder holder from North Afghanistan (1-22) is another object still made in the traditional way. It, too, uses a modification of a textile process to form a textural covering over a clay object. The clay is first covered with a purple fabric and then wound tightly with a continuous wire threaded through an expanded wire coil. The coils are not actually interlocked, although they give that appearance. Rather, they are placed in close enough proximity to each other so that they cannot shift their position and they hold as firmly as if worked in an interlocked textile structure.

Quite similar in appearance to the near Eastern pieces is an intriguing traditional piece of coin silver from Nigeria, made in the late nineteenth or early twentieth century. The chain is spool-knitted on seven pins with a double row of loops that creates a dense structure (1-23). I have seen only two or three African chains of this type in which the attribution is certain, although such work might be more widespread.

1-22. A powder horn from North Afghanistan uses a single wire threaded through a continuous coil to wrap a clay form in a textile-like structure. Courtesy of Jack Lenor Larsen.

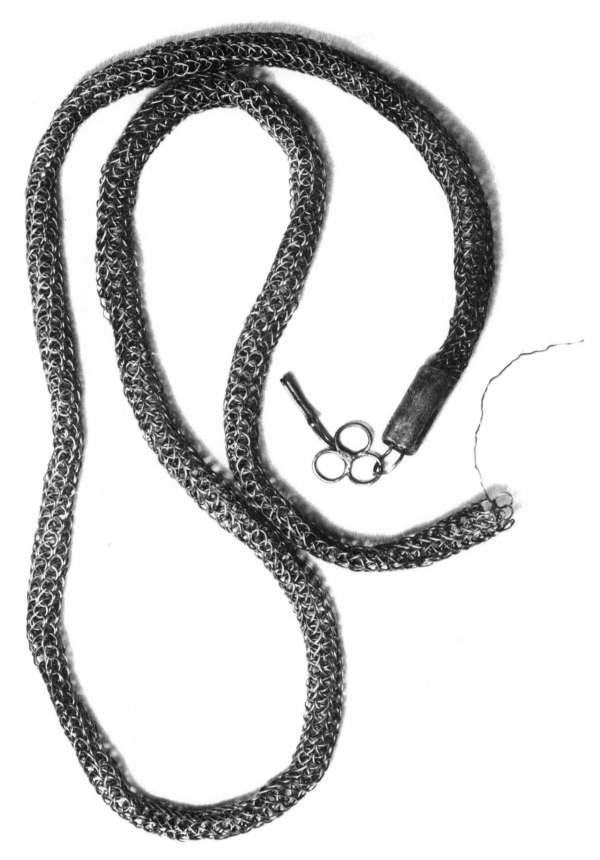

1-23. A knitted neck chain in coin silver from Nigeria, late nineteenth century.

The last two historical examples are extremely fine
wire bags made in Germany or France about 1830.
Both bags use a zigzag chain-link structure throughout
to produce a fabric which is flexible in only one direction—the interlinking rows shift up and down, com-
pressing and expanding the form. The small bag
(1-24a) is somewhat flexible even though the edges
are laced with a supplementary wire in a continuous
overcast stitch, which tends to stiffen them. The ornamentation at the top and bottom (1-24b) is of particular
interest, consisting entirely of fine wire coiled over
heavier wires which are then fashioned into edging
and hanging elements for a strong visual effect. The
large bag (1-25) uses the same basic interlinking structure but loses its fabric quality almost completely by
the addition of stiff rods at the edges and by the super-
imposition of a diagonal pattern in a reverse twist.
Individually constructed and shaped sections are sewn
together with continuous wire coils which incorporate
the stiff rods at the edges. The only true metalworking
techniques are found in the stamped sheet-metal ornaments and the catch, which are riveted in place, and
in the chains that form the handle. Interlinked structures have recently become popular with craftsmen,
and adaptations can be seen in Chapter 7.

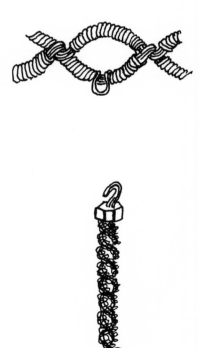

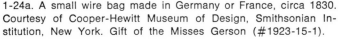

1-24a. A small wire bag made in Germany or France, circa 1830.
Courtesy of Cooper-Hewitt Museum of Design, Smithsonian Institution, New York. Gift of the Misses Gerson (#1923-15-1).

1-24b. Ornamentation on the bag. *From top:* details of basic
structure, upper edge, and hanging elements.

1-25. A wire bag from Germany or France, circa 1830. Courtesy of Cooper-Hewitt Museum of Design, Smithsonian Institution, New York. Gift of Miss Diana del Monte (#1920-5-1).

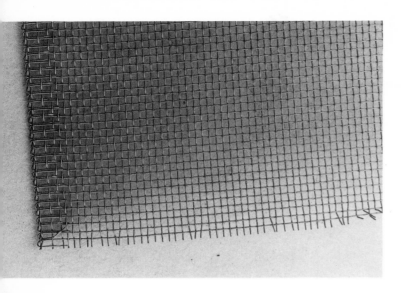

1-26. A section of ordinary window screening.

Our own twentieth-century culture has created a number of industrially produced objects using metal in textile structures, some purely functional and others meant to be decorative. In the purely functional realm are things like window screening (1-26), chain-link fencing, a whole range of industrial tubing made of braided wire (1-27) and used for battery cables and the housing of multiple circuit wires, as well as domestic items as diverse as a wire vegetable basket (1-28) or the inside of a plastic car seat (1-29). The vegetable basket is interlinked so it can collapse and expand constantly without damage to the structure or the form. The car seat is a continuous coil of hardened steel wire fashioned into a spiral and held in place by a single wire that interlocks the adjacent coils to each other. Since this is the interior structure of an object which is meant to be stationary, it is held in an expanded state by a rigid frame around the perimeter.

1-27a,b,c. Three samples of industrial copper tubing constructed in a braided technique using multiple strands of copper wire as the structural element. The tubing is available in a wide range of sizes.

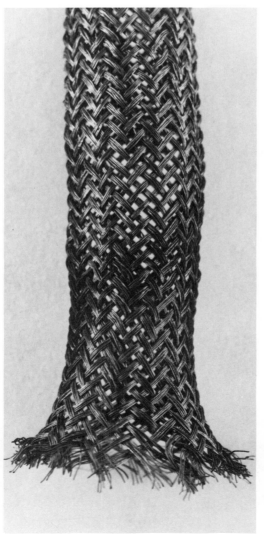

1-28. A vegetable basket constructed in an interlinking technique that allows it to collapse and expand its form.

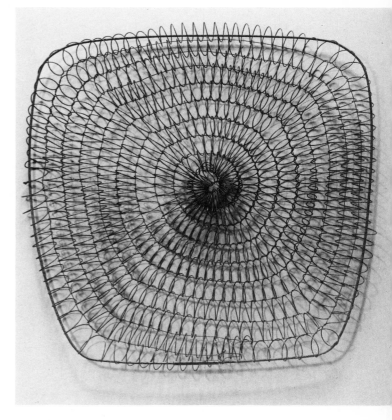

1-29. The inside structure of a car seat.

Of the decorative items, I have chosen two which are constructed completely with textile processes using non-precious metals. They are not pretentious either in material or form, and the decision to use metal in a textile structure was probably based more upon practicality than aesthetics. The candle holder (1-30) is one of a pair fashioned totally by the weaving process in a combination of brass and copper wire. Both elements are double wires; a heavier gauge is used for the warp (.85 millimeters) and a smaller gauge for the weft (.55 millimeters). There are no solder joints or other metalworking techniques used—the entire form is achieved through the weaving process. The lamp shade (1-31) probably is a variation on the construction of a rattan or reed basket, and its overall design is certainly reminiscent of basket forms. Two widths of brass sheet are used, the larger to construct the openwork area in the center and the smaller to weave the dense bands at top and bottom. Again, no metal technology is used in the piece, which is begun, formed, and ended totally with textile processes. For actual use it is lined with silk or thin cotton fabric to filter the light.

Wherever one looks, the examples of textile techniques in metal, both historical and contemporary, keep multiplying. This survey barely scratches the surface. Great numbers of intriguing and provocative metal fabrications wait to be found either in distant lands and centuries or in the hands of contemporary craftsmen. Once one is aware of the techniques, every new example that comes to light is an exciting surprise.

1-30. A candle holder made of brass and copper wire in a completely woven structure, 7½ inch total diameter, 1″ center cylinder, twentieth century.

1-31. A lamp shade totally constructed with strips of brass in a combination of basketry techniques, twentieth century.

2/Basic Materials, Tools, and Processes

MATERIALS

Textile constructions involve much bending, looping, and interlacing, usually carried out under tension as in weaving or with the constant tugging and pulling necessary in crochet or knitting. String or yarn is the most flexible of materials, and, when using metal in its place, the craftsman must always consider malleability and strength in determining the suitability of material to process. The strip of metal used for weaving may not be malleable enough for crochet, and a wire thin and flexible enough for knitting might not have strength enough to withstand the tension put on it in sprang.

Malleability refers to the manner in which a metal can be manipulated and how long it can be manipulated before it breaks. Fine, or 24-karat, gold is the most malleable of metals. It has an almost infinite capacity for being hammered, stretched, bent, twisted, and pressed without breaking. At the other end of the scale is something like a wire coat hanger. Its base metal is very difficult to bend and almost impossible to stretch, hammer, or twist because it is so hard and brittle. It has a very short capacity for being manipulated with tools or the hands, so that after just a few twists, it will fatigue and fracture. In between these two extremes lies a vast range of metals in sheet or wire form, each with its own level of malleability.

In metals, strength means the ability to withstand stress of many kinds. Some metals break easily if bent suddenly; others gradually weaken as they are compressed and expanded, or slowly deteriorate due to atmospheric conditions or a natural process of aging. Strength also has something to do with the way the metal will retain its shape. Some metals that can be fashioned into a form while pinned to a board or nailed to a frame will not be able to maintain that form unsupported. The wire used to make chain-link fencing is similar in hardness and stiffness to coat-hanger wire, yet it is suitable for the fence's single interlinking

in a zigzag pattern (similar to sprang) and has sufficient strength and durability to give it a long life in any climate. In a smaller gauge, the same wire might be suitable for a basket structure, since it creates a self-supporting form, but it would not bend enough for crochet or knitting.

In the use of sheet metal or flat ribbons of metal the same relationships exist between the properties of the material, the technique to be used, and the final product. Sheet metal generally needs to be in a softened state if it is to be manipulated very much in any manner. Most nonprecious metals are supplied in a hard or semi-hard state unless otherwise specified, in which case they can be softened through the use of heat in a process called annealing.

The thickness, or gauge, of these wires and sheets is of the greatest importance in determining the working properties of the metal. The more malleable a metal, the finer it can be drawn or rolled. Gold, for example, can be made into the finest of wires and very thin gold leaf. Metals which are referred to as *foils* or *shim stock* are very thin gauges ranging from the ordinary household aluminum foil, which is only paper weight, to copper roof flashing, which is perhaps as heavy as 24 gauge. The lower the gauge number, the thicker the metal is. Ordinary sheet stock in almost all metals ranges from these foil-like weights (40 gauge and thinner) to eighth-inch thicknesses (8 gauge) and even heavier.

Most textile techniques utilize the thin end of the range, which is more flexible for working. They may be identified by a variety of measurement systems that need to be understood when ordering supplies. In the United States the common measurement (as used above) for most wire and for *precious* metals in sheet form is the Brown and Sharpe Gauge. It uses simple digits ranging from 1 to 40, with the thickness diminishing as the numbers grow larger. *Nonprecious* metals in

sheet form are sometimes measured by ounce weight per square foot of the material; naturally, as the weight increases (larger numbers) thickness also increases. This means that the numbers derived from the weight system work in the opposite direction to those in the Brown and Sharpe Gauge, so it is necessary to know which measurement system is being used when buying sheet metal.

A less confusing system is that of actual measurement in thousandths of an inch, which is standard for all wire and sheet metal. In most of the world the metric system is used exclusively, greatly simplifying the whole process. A comparative table of measurements is included in the Appendix for quick reference, and it should be carefully checked when ordering or buying metals from unfamiliar sources to insure receiving the correct thicknesses.

2-1. Metal in wire form. *From left to right:* 24-gauge coated copper, 28-gauge copper, 24-gauge uncoated copper, 30-gauge fine silver (below), small spool of 38-gauge brass wire (above), 30-gauge 18-karat gold wire.

2-2. Metal in sheet form. *From back to front:* copper roof flashing (12 inches wide and 28 gauge); brass shim stock (6 inches wide and 32 gauge); fine silver (2 inches wide and 26 gauge); sterling silver, roll at left (6 inches wide and 30 gauge).

Following pages

2. A warp of brass is woven with copper foil weft elements; the ends are fastened with silver rivets.

3. Sterling-silver sheet and fine-silver wire used as warp, with some elements oxidized. The weft is a continuous strand of fine-silver wire.

4. Bobbin lace in a Labyrinth pattern, constructed of two different gauges of fine-silver wire.

5. Two colors of gold, 18-karat green gold and 24-karat (yellow) gold, are used in the warp. The weft is an alternation of the same elements with the addition of 18-karat gold wires that have been hammered at the ends.

6. A braided construction uses 24 strands of wire in three metals, silver, copper, and red-coated copper.

7. A necklace of hairpin crochet made by joining seven crocheted bands of fine-silver and coated copper wire.

8. A knitted construction uses a double element of fine-silver wire and coated copper wire in the heavy border area and a single fine-silver wire for the lacy center section.

9. A crocheted construction uses multiple colors of coated copper wire in a variety of dense and open stitches. Detail of a belt by Christine Brown.

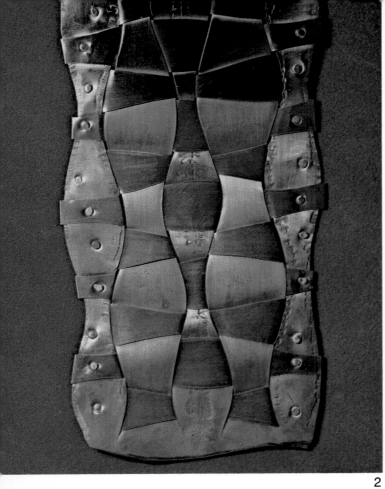
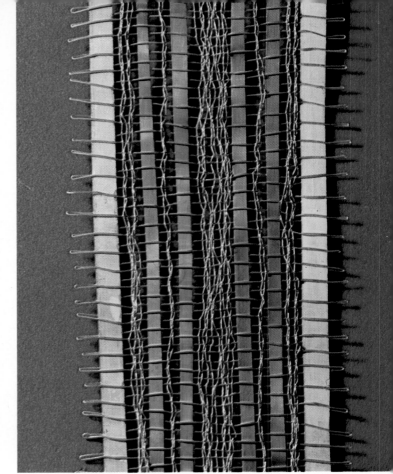

2   3

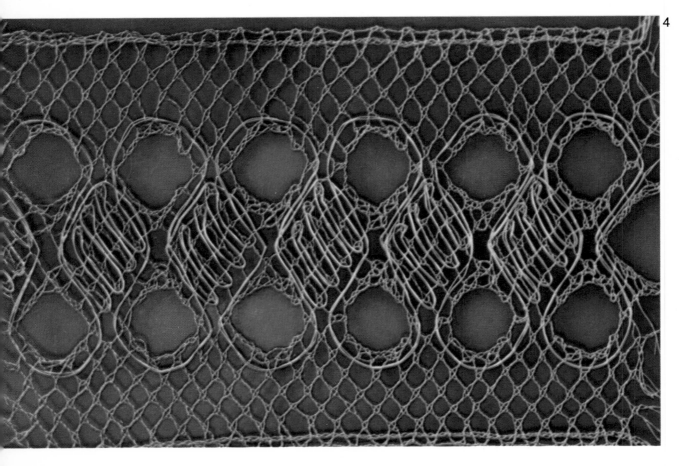

4

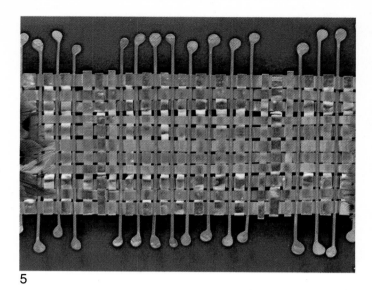

5

8

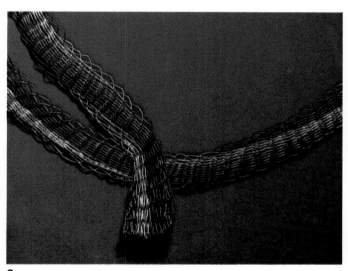

6

9

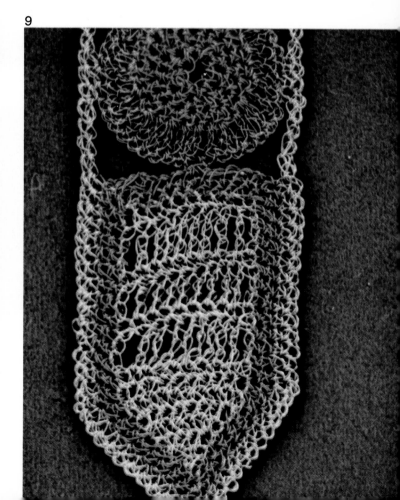

7

METALS FOR TEXTILE TECHNIQUES

Gold, silver, copper, brass, and pewter are commonly used in making jewelry, as can be seen in many of the pieces illustrated throughout the book. Especially in wire form, they are malleable enough to be worked almost without the aid of tools. Although brass is a little harder to work, thin-gauge sheets can be cut easily into strips for various uses including weaving and caning.

Many other metals can be used in textile processes, ferrous as well as nonferrous. Both groups may be used together in a single piece provided that any heating and cleaning operations are compatible with all of the metals included. In this regard, it is important to note differences in the melting point of the metal (see Appendix) and its alloys, and the differences in reaction to heat and chemicals used in metalworking (see the processes described later in this chapter).

GOLD

Gold wire and sheet is available in a variety of gauges and alloys. Fine gold is too soft and too expensive for most purposes unless it is interlaced with a stiffer material to help maintain the structure of the piece. An alloy of 18-karat gold is sufficiently malleable to be used successfully in most textile processes, and if the gauge is thin enough, does not have to be annealed. White gold (alloyed with platinum, palladium, or nickel) and green gold (alloyed with silver and copper or silver and cadmium) tend to be much harder than a comparable karat in yellow gold, and a thinner gauge of these alloys will be necessary to obtain the same flexibility. Gold is an excellent choice for textile techniques that require constant bending and twisting since it is so malleable.

Processes which use delicate wire generally are of the single-element type; therefore a continuous length of wire is highly desirable in order to avoid the problem of joinings and splices within the structure. In heavier gauges gold wire comes coiled. Having finer-gauge wire on a spool is also a great advantage in keeping the material free of tangles, kinks, and twists, which can prove troublesome. Both flat and round wire can be bought already annealed and spooled. Since most craftsmen are not equipped to make sufficient lengths of such fine wire, it is advisable to order the material already spooled even though there is a refiner's charge for the special processing, and such special orders normally require a minimum quantity of five ounces. Five ounces of 30-gauge wire provides sufficient wire to produce several pieces. Another excellent alternative is to use cloisonné wire of 18- or 24-karat gold, which can be purchased from metal or enamel suppliers in a limited range of widths and thicknesses (from 16 to 30 gauge) without special ordering, and is immediately usable for all textile processes.

Since most work done in gold will probably be small scale, it is important that the material be of a thin gauge so that it is in proper proportion to the finished piece. The knitted gold pieces illustrated here are all executed in 18-karat gold wire of 30 gauge (.010) and required no annealing either during or after fabrication. Even finer 18-karat gold wire of 32 or 34 gauge (.007 or .006) can be used for crocheting techniques that provide structural strength through the build-up of density rather than using only the strength of the material for support.

Gold sheet, too, must be thin enough to be malleable and suit the scale of the piece. Any size can be ordered but thinner gauges generally cost more. If a rolling mill is not available, it is necessary to order sheet in the gauge needed for a particular project, for which there will be additional refining charges. As illustrated here in the woven bands on a ring, it is possible to use relatively inflexible sheet gold (as low as 14 karat) for weaving strips if the sheet is rolled down to 30 gauge and thoroughly annealed before use. In thin gauges, it is a simple matter to cut sheet gold into strips with an ordinary pair of sharp scissors. As an alternative, strips of gold can be created by rolling down round wire into flat, even widths, and then annealing. Some cloisonné wire may also be suitable.

SILVER

Two types of silver can be used readily in textile processes—fine silver and sterling silver. Those techniques which use a single continuous element, such as knitting or crochet, and those which use a single set of elements, such as braiding or lace, are much more successfully accomplished in fine-silver wire, because it does not work harden as quickly as sterling under the constant manipulation these processes require. After a piece is finished, it may have to be manipulated frequently in normal use, as, for example, in a piece of jewelry that is constantly opened and closed. Fine silver is very soft in its original state and requires a minimum of annealing, and should be used in any piece which involves many complex movements either during or after fabrication.

Sterling, which is a particular alloy of silver with copper (925:75), is much better able to hold its shape than fine silver but will not tolerate constant movement. Eventually it will begin to fracture at points of greatest stress. Sterling is preferred in most jewelry and hollow ware because of its greater strength, and can be easily softened by annealing when necessary, although it requires pickling afterwards (see below). Because it is often difficult or impossible to anneal frequently while work is in progress, sterling should be used only in techniques such as weaving and interlinking, where the metal is subjected to minimum stress and the minimum number of sudden or drastic changes in direction. Any technique which uses more than a single set of elements can be accomplished well in sterling, although its natural springiness will resist being pushed into a too closely interlaced structure.

Sterling and fine silver are both readily available from dealers and distributors of precious metals. Sterling tends to be kept in stock in a greater range of sizes (from 8 to 30 gauge) than is fine silver, which frequently has to be made to order in the smaller gauges. As with gold, silver sheet is easily rolled down to thinner gauges using a rolling mill, which greatly reduces the variety of stock that is necessary to keep on hand in a private workshop. The sheets ordinarily come in six- or ten-inch widths cut to any length and in the standard 18-, 20-, 22-, and 24-gauge thicknesses.

Wire in the very thin gauges presents the same problem in silver as it does in gold—most small workshops are not equipped to make large unbroken lengths of small-diameter wire. If short lengths of several feet or yards are required, 12-gauge round wire can be drawn through a hand drawplate (see the sections on tools and processes later in this chapter). The longer lengths are best ordered through a dealer or directly from a manufacturer, with the specification that they be spooled rather than coiled. As previously mentioned, spooled wire is easier to work with in continuous-element techniques. Most dealers have minimum orders of five- or ten-ounce lots, while some manufacturers require a 50-ounce minimum before a special order will be accepted. A 50-ounce spool of 26-, 28-, or 30-gauge wire is an enormous supply for a craftsman, and will last through a number of projects. Silver cloisonné wire, which is a thin flat wire produced in a small range of sizes of fine silver (16 to 30 gauge) is available from both metal distributors and enamel suppliers and may be purchased in small quantities.

COPPER

Commercial copper is an almost pure metal that is extremely malleable in its natural state. It behaves much more like fine silver than like sterling, and is an ideal metal for use in all textile techniques.

Copper is the most readily available, generally the least expensive (although prices of all metals fluctuate, sometimes rapidly), and the most versatile material for use in textile processes. Because of its softness and extreme flexibility (only 24-karat gold is as flexible), the wire is ideal for all textile techniques from knitting, crochet, and braiding to weaving with cards or on a loom. Strips of sheet are good for weaving, basketry, and braiding.

Copper sheet is manufactured in three states, hard, half-hard, and soft, and by two different processes, hot-rolled and cold-rolled annealed. Hardness has to do with both annealing and rolling. Copper which is hot-rolled is soft, while cold-rolled copper must be ordered in an annealed condition or else it must be annealed before using. Since cold-rolled metal has a smoother and a more nearly perfect surface, it is the preferred one to order, if there is a choice. Soft or annealed copper is available in sheets of three feet by eight feet, as well as in rolls of eight-inch to eighteen-inch widths. For most textile techniques, the rolled metal sheet, which comes in gauges from 18 to 36, is especially suitable since it is easily handled, cut, and stored. The thinner gauges (24 to 36) can be cut with regular scissors.

Copper referred to as *foil* and *roof-flashing* is of the rolled type, and is available at building suppliers as well as in hobby shops. Wherever possible, it is advisable to purchase sheet copper from a sheet-metal dealer or from a supplier to the building trades, since the price will be much more favorable than that of most hobby and craft suppliers, who sell materials in small prepackaged parcels.

Copper wire can be purchased on spools in a wide range of sizes, from 18 to 40 gauge, from a number of different sources. Copper wire of very fine gauge is used in electronic circuitry and is available at electronic supply dealers, radio and hi-fi shops, and frequently at salvage yards and discount surplus dealers. Wire made for circuitry is usually coated with a varnish which retards tarnishing and gives distinctive color. The varnish is so well bonded to the metal that it will not crack or peel when manipulated in any textile process. The varnish will burn off if the wire is annealed, it will flake off if the wire is rolled through a mill, and it can be removed chemically if the finished piece is to be plated with another metal (see below). In all other circumstances the varnish will remain and, since it comes in a variety of colors, can be used to great advantage in creating color patterns in the finished piece. Varnished wire comes in such brilliant colors as bright red and emerald green and in a number of subtle copper tones ranging from dark brown to light gold.

Copper wire is used for other purposes besides circuitry and is frequently available in very large quantities at a minimal cost. Motor-armature wire for machinery is sold on large spools in a variety of gauges (from approximately 18 to 40) and a subtle range of copper tones. This wire tends to be less expensive because of the larger quantity being sold on the spools and is usually priced by the pound. A single spool may hold as much as ten pounds. Small spools of one-quarter to one-half pound are adequate for most small projects and samples; a large woven piece might use several pounds. Uncoated copper wire also comes on smaller spools weighing one-quarter to one-half pound each and is available from electrical shops as well as radio and electronics suppliers. Manufacturers of this type of wire can be contacted directly for a listing of local distributors.

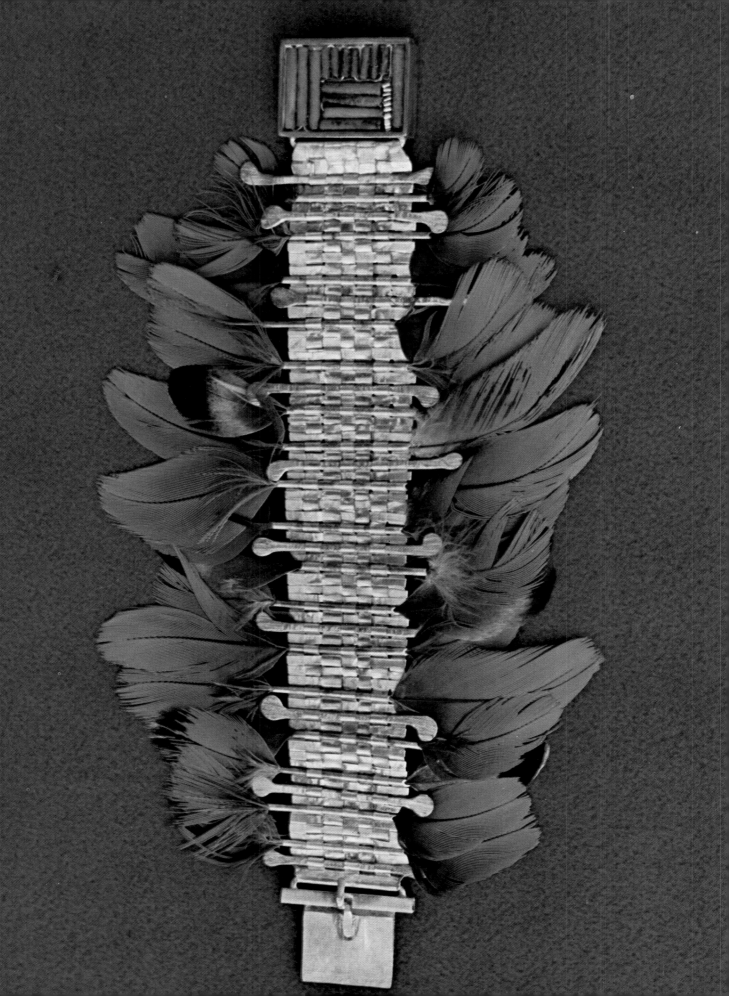

BRASS

Brass is an alloy of copper and zinc and comes in a variety of tones from gold to pale yellow. Yellow brass, which has the greatest percentage of zinc in the alloy, is the most widely used since it is the most malleable of the brass alloys, and has the lowest annealing temperature. Brass in general is not as soft or as malleable as copper, and must be pickled after annealing (see the section on processes). In thin gauges, it performs very well in textile processes. The sheets, which are sold in a hard state and are too springy to be used effectively until annealed, are easily cut into strips for weaving, caning, or other processes which use ribbons of metal. It is good material for work where strong self-supporting structures are needed.

Sheet brass is more difficult to locate than copper, although it is manufactured in the same size sheets and rolls. The most common form of thin-gauge brass sheet is called *shim stock,* and is available in rolled widths from two to twelve inches. It can generally be obtained from plumbing or building suppliers as well as hobby shops.

Brass wire is not as widely used in industry as copper and cannot be located easily in large quantities. Very thin gauges (perhaps 36 or 40) can be found on small spools similar to sewing thread in craft and hobby shops and from floral display dealers. Such thin wire can be used without annealing but must be handled gently to avoid breakage since it is not very strong. Heavier gauges (16, 18, 20, 24) are generally available in hardware and home-repair shops. It is packaged either in coils or on small wooden spools weighing approximately two to four ounces, and is in a soft state, ready for use.

PEWTER

Pewter, an antimony-copper alloy of tin, is extremely malleable, being softer than brass, bronze, or precious metals of comparable gauge. As it never work hardens and so does not have to be annealed, it would be an ideal material for using in single-element processes such as crochet or knitting if it came in small-gauge wire. The wire is difficult to obtain and generally comes in large diameters of about 14 gauge up to one-and-a-half inches. It would be suitable for a piece on a fairly large scale.

Like copper, pewter does not hold its shape well in thin gauges and is most effectively used for structure when in combination with other metals. With copper and brass it provides an interesting color change and a very dimensional effect as it moves over and under the harder metal elements. Combination pieces with pewter cannot be subjected to high heat, as pewter has a very low melting point, so any annealing of the accompanying metals must be done prior to the construction process.

Pewter is available in sheet form in gauges as thick as 14 down to foil of about 36 gauge. The sheets come in many sizes, usually 12 or 16 by 24 inches, and can be cut with ordinary shears (24 to 36 gauge thicknesses). They can be obtained from craft shops in small quantities but are more economical when purchased from the rolling companies if the usually required minimum order of twenty-five pounds for sheet and five pounds for wire can be accommodated.

Opposite page
10. Woven bracelet in fine silver with blue macaw feathers. The clasp is constructed of sterling silver and inlaid with Egyptian mummy beads (faience).

OTHER METALS

Many other metals may be used if they are available in the forms and sizes needed. Aluminum sheet could be used for large-scale architectural pieces because it is lightweight, malleable, highly resistant to corrosion, and may be colored by anodizing, which is similar to the electroplating process described later in the chapter. It is important to check the softness of aluminum when it is purchased since there are a large number of different alloys made, some of them extremely tough and hard with a high degree of tensile strength.

Stainless steel is produced in both sheet and wire form although the wire is probably more suitable for use in textile processes. Stainless-steel nautical wire is made for use on the rigging of sailboats and may be found in single or multiple strands of small gauge.

Salvage yards, scrap metal dealers, and surplus dealers are probably the best sources for all non-precious metals when small quantities are needed for experimentation or for small-scale projects. Most manufacturers and distributors are unequipped to deal with small orders, and will frequently add exorbitant "lot" charges for the inconvenience of filling such orders.

The following chart of metals and textile techniques may help the beginner to decide what is most suitable for an intended project. The suggestions are by no means the only workable ones, and they are provided with the hope that further experimentation will be stimulated. Additional suggestions will be found in the Appendix.

Quick Reference Chart for Materials and Textile Techniques

XX recommended X possible	GOLD			SILVER		COPPER	BRASS	PEWTER	OTHER METALS
	24k	18k	14k	Fine	Sterling				
WEAVING									
sheet	X	XX	XX	XX	XX	XX	XX	XX	aluminum, steel
wire	X	XX	XX	XX	X	XX	XX	XX	
KNITTING									
needle	X	XX	X	XX		XX	X		stainless steel, iron
spool		XX		XX		XX			binding wire
CROCHET	X	XX	X	XX		XX	X		stainless steel
BRAIDING	X	XX	XX	XX	X	XX	X	X	
(also cloisonné wire)—									
INTERLINKING	X	XX	XX	XX	XX	XX	XX		stainless steel
SPRANG		XX	XX	XX	XX	XX	XX		stainless steel
BOBBIN LACE	X	XX	XX	XX	X	XX			stainless steel
BASKETRY									
Plaiting	X	XX	XX	XX	X	XX	XX	X	aluminum
Twining	X	XX	X	XX		XX	XX	X	
Coiling	X	XX	X	XX	X	XX	XX		

TOOLS

Although textile constructions in metal use basically the same tools and implements as yarn, there are a few metalworking tools that are essential for cutting, bending, stretching, and forming the metals. Some of these are sufficiently common to be found in local hardware outlets, while others are specifically jewelers' tools available only from more specialized sources.

CUTTING TOOLS

1. Scissors. Large paper-cutting scissors, six, eight, or ten inches long, are most satisfactory for cutting sheet metal of 24 gauge and thinner.
2. Plate Shears. Metal-cutting shears are intended for cutting sheet up to 18 gauge. These may be purchased in a variety of sizes and with straight (see 2-3) or curved blades, which are useful for cutting curved or circular forms.
3. Wire cutters. These are used specifically for clipping wire and are superior to shears for this purpose. They come in a wide range of sizes from four to eight inches, depending upon the dimension of wire to be cut. For most of the soft, fine-gauge wires, a small pair of four or four-and-a-half inches is adequate, while a larger and heavier version is necessary for wires of 10 gauge or more. There are two distinctly different shapes: end-cutting and diagonal cutting, with personal preference the only determining factor (see 2-3). Both types clip the wire in a wedge shape that may require hammering, beading, or filing to produce a finished ending.

2-3. Hand tools for cutting and bending. *Top row, left to right:* diagonal-cutting wire nippers, round-nose pliers, chain-nose pliers; *bottom row, left to right:* end-cutting wire nippers, flat-nose pliers, chain-nose pliers (fine point); *far right:* plate-cutting shears with straight blades.

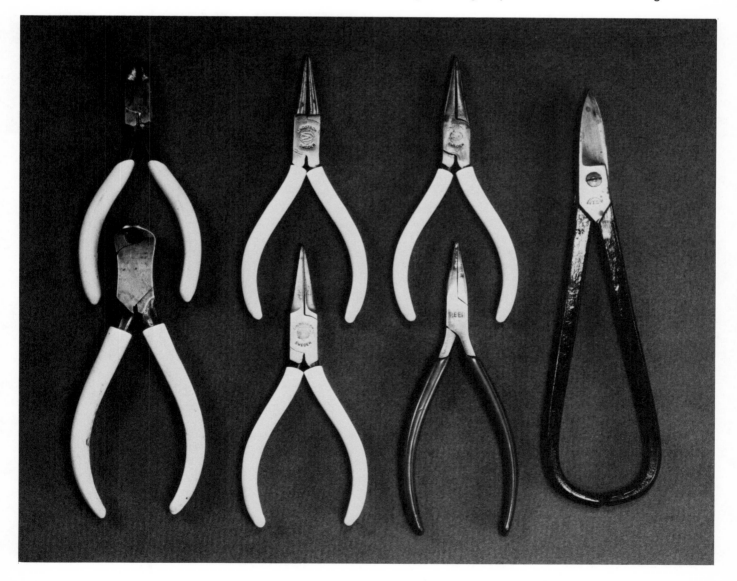

BENDING AND STRETCHING TOOLS

Wires of all gauges can be curved, coiled, bent, twisted, and shaped by jewelers' pliers, which have smooth, rather than serrated, jaws and mar the wire as little as possible. A variety of shapes are available including chain-nose, round-nose, flat-nose, half-round and forming, of which the first two are the most essential for use in textile constructions (see 2-3). All pliers can be found in sizes from three to six inches, with four to five inch lengths the most common and easily used.

The diameter of wires can be made smaller with the use of a drawplate and a pair of draw tongs (2-4). The drawplate is a quarter-inch thick piece of hardened steel with a series of graduated holes tapered to reduce the cross section of wire as it is pulled through successive holes. Drawplates are made with holes of round, oval, square, rectangular, half-round and other cross-sections so that the shape as well as the size of the wire can be changed by the drawing process. The tongs used for drawing are especially designed for the purpose, with heavy serrated jaws for holding the wire firmly and a hooked handle to facilitate gripping while pulling the wire through the holes. The drawplate is held upright in a firmly fixed vise since the wire must be pulled hard and smoothly.

Sheet metal can be stretched thinner and longer by the use of a hand rolling mill (2-5). This is a machine with two highly-polished rolls of hard steel which are geared for easy turning. The rolls are kept parallel, but the opening between them can be adjusted from about a quarter inch to .010 inches and less. The metal is stretched only in the direction in which it passes between the rolls (generally its length, not its width), so the width of the rolls determines the maximum width of the sheet to be rolled. Most hand mills range from two inches to five inches in width.

2-5. A hand rolling mill with three-inch rollers and double handles.

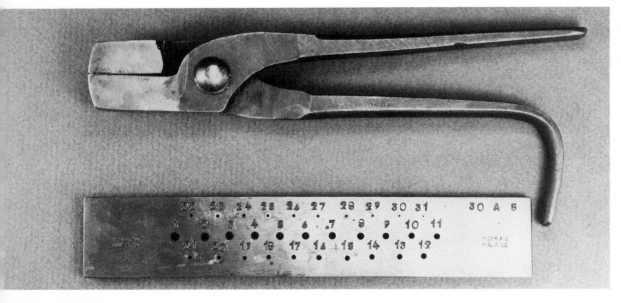

2-4. Implements for drawing wire. At top: a pair of draw tongs; at bottom: a drawplate for round wire.

FORMING TOOLS

Wire is easily formed in the fingers or with a pair of pliers. Sheet metal, however, can be marred easily by the edge of the jaws of pliers and is not easy to form with the fingers, except when it is a very thin gauge. A few basic hammers and a rawhide or plastic mallet provide many possibilities for forming and texturing sheet metal of any thickness. Mallets are hammers made of materials softer than metal, which can be used to flatten or to form any sheet without permanently damaging the surface. They may be made of wood, horn, fiber, plastic, or rawhide leather and are available in a wide range of sizes and shapes. A most useful one is the rawhide version pictured in 2-6. The head is one-and-a-half inches in diameter, flat on both sides, and the handle is ten inches long.

Steel hammers serve a variety of functions. They can flatten wire and sheet; they can stretch both wire and sheet by altering the thickness; they can produce curved and domed forms by selective stretching; they can create textural surfaces by the multiplication and overlapping of marks that are the natural result of the hammer striking the surface. There are hundreds of hammer shapes used by the silversmith, but two or three very simple ones are more than adequate to accomplish all functions related to textile processes.

2-6. Hammers for forming sheet and wire. *From bottom to top:* rawhide mallet, ball-pein hammer (4 oz.), and planishing hammer (2 oz.).

1. Ball pein hammers. These hammers have a flat face on one end and a rounded or hemispheric form, the pein, on the other end. They are available in hardware stores in a variety of weights from two ounces to several pounds, with a four-ounce size being the most suitable for small-scale work (see 2-6). The flat face has a very slightly domed surface which should be retained since it prevents any accidental striking of the edge of the hammer on the sheet metal, an error that leaves a disfiguring mark on the surface. Both hammer faces should be kept highly polished since all blemishes will be repeated on the surface of the softer metals being hammered. Polishing can be done with diminishing grades of emery paper, followed by machine buffing with a Crayatex wheel or by hand polishing with a felt buff impregnated with pumice or white diamond polishing compound.

2. Planishing hammers. These hammers have two flat faces with one end slightly domed for working on flat surfaces, and, as they are used specifically by jewelers and silversmiths, will be available only from specialized suppliers to these trades or from craft-tool dealers. The hammers come in weights ranging from two ounces to sixteen ounces. A two-ounce hammer is excellent for small, detail work on both wire and sheet (see 2-6), while a six-ounce hammer is more practical for general forming, stretching, and surfacing of sheet metal. It is imperative that these hammers be kept free of all surface blemishes and that they be highly polished for the most effective usage.

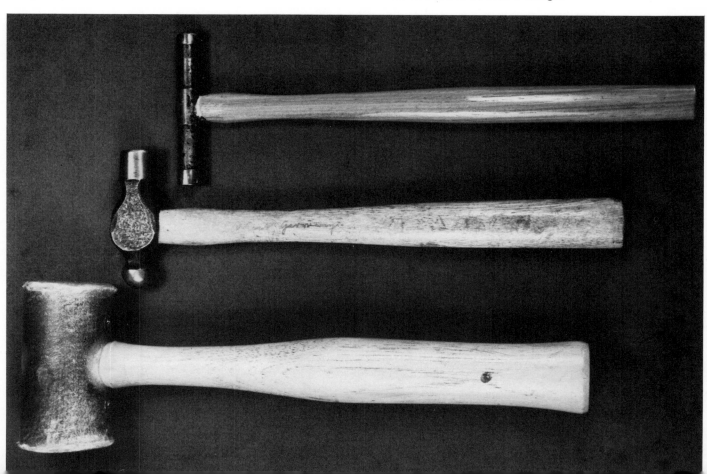

3. Rivet hammers. Generally smaller than other hammers in both size and weight, these have one wedge-shaped end used for spreading or stretching rivets and one small, flat, or slightly domed, round face for smoothing and rounding rivet heads. These hammers may have a double head only one-and-a-quarter inches in length and weighing as little as one-half ounce (usually called watchmakers' hammers). A two-ounce version may be more useful (2-7).

In addition to hammers, some simple assortment of surfaces are necessary for the forming process—soft surfaces, which allow the metal sheet to sink into relief, and hard surfaces, which permit stretching, refining, and texturing of the sheet metal. In the first category are sandbags made of canvas and filled with sand (2-7), lead blocks of one-half-inch to three-inch thicknesses (2-8), and wood forms and blocks made of hardwood of any dimension, from a small square to a large tree stump. In the second group are small square blocks of steel, anvils (2-7, 2-8), round steel mandrels in the shape of long fingers, and dome-shaped steel stakes and anvil heads (2-8) all used as forming implements and refining surfaces against which softer-than-steel metals can be hammered.

2-7. Forming tools. A 2-oz. rivet hammer rests on a sandbag next to a steel anvil, 4 x 4 x 1½ inches.

2-8. Additional forming tools. *From left to right:* a thick lead block 4 x 4 x 1 inch, a T-stake with three different forming surfaces, a flat-top rectangular anvil head, a mushroom stake, and two small anvil heads.

MISCELLANEOUS TOOLS

A few other common hand tools will be useful when dealing with metal before and after it is used in textile constructions.

1. Files are used for shaping metals and for finishing the edges of sheet metal and the clipped ends of wire. One large flat file (usually called a hand file) eight inches long, of a Number 1 single cut, one six-inch half-round file of a Number 2 double cut, and several five-inch or six-inch needle files of Number 2 cut in different cross sections (round, square, knife, warding) will provide an adequate assortment for almost all of the metalwork connected with textile processes (2-9).

2-9. Files. *From top to bottom:* a half-round #2 cut file, seven needle files in a variety of cross sections (triangular, square, warding, half-round, knife, flat, round), and a large flat #2 single-cut file.

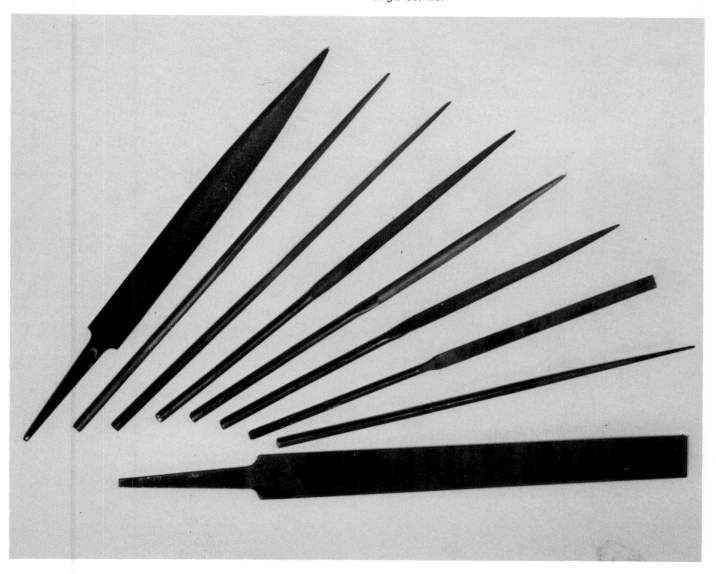

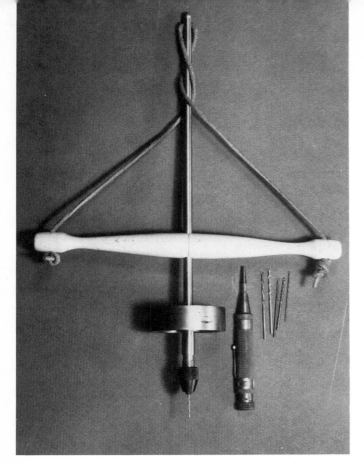

2-10. Tools for drilling holes. *From left to right:* pump drill, automatic center punch, and an assortment of twist drills in small sizes.

2. A hand drill will be necessary for making the holes for rivets or for any other structural or decorative details. The most common is the egg-beater type available in any hardware store, with a chuck which accommodates twist drills of one-quarter inch diameter and less. Another type is the pump drill available from jewelry-tool suppliers, which has the advantage of single-handed operation (2-10). Both types use twist drills made of carbon steel with a shank which is inserted into the drill chuck. These are available in a wide range of sizes, with the most useful for jewelry work being in the Number 50-65 range. A center punch is also necessary for the drilling operation. It is used with a hammer to produce a small indentation in the metal that holds the drill in place and ensures accuracy. A large nail may be used instead, or a more elaborate spring-loaded automatic center punch (2-10) that is operated without the use of a hammer.

3. A wooden clamp with a wedge insert is an excellent holding device for filing, coiling, and many delicate finishing operations. Its leather-covered jaws prevent any marring of the metal surface (2-11).

4. A pair of pointed jewelers' tweezers are useful for handling and holding small objects and bits of metal. They are essential for any soldering operation, and may also be used for holding wire during the beading process.

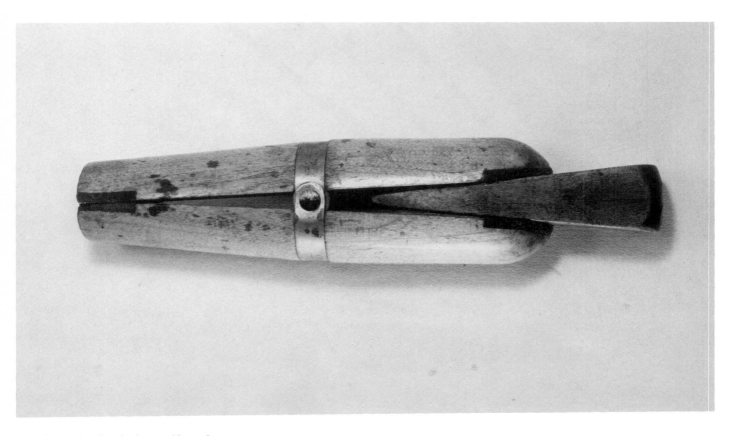

2-11. A wooden hand clamp with wedge.

SHOP EQUIPMENT

Although textile processes in metal can be done without any metalworking equipment, the number of embellishments and refinements that can be made is greatly increased if a few shop tools are available. There is also a far greater possibility for preparing and treating the metals prior to construction if some very basic shop equipment is on hand.

1. A torch for annealing and soldering (see next section). A very inexpensive heat source is the small propane gas tank sold for camping stoves and a variety of household repairs. It may be purchased with interchangeable tips which provide a variety of flame sizes for different operations. A more expensive and elaborate outfit uses compressed acetylene (usually Prest-O-Lite) and requires the separate purchase of a tank, regulators, torch handle, and interchangeable tips (2-12). These are generally available from welding suppliers and amount to a sizeable investment. Oxy-acetylene units for welding and brazing may be used but are not recommended because the heat is far in excess of what is necessary for nonferrous metals, and may prove damaging when dealing with small-scale materials.

2. A turntable for soldering and annealing. Any simple metal turntable may be used if the top is covered with an asbestos material. Regular annealing pans are usually filled with carborundum grains or pumice stones for greater heat reflection. Simple fire bricks or blocks of charcoal are quite satisfactory as surfaces for heating, although they lack the advantage of the more even distribution of heat provided by the turning operation,

3. A flame-proof pot for pickling solutions (see next section). These must be of a glass or ceramic material since the acid solutions react with most metal containers, lead being an exception. Pickling solutions are most effective when hot, so it is best if the container can be placed over an electric burner or low gas flame when in use.

A recommended list of basic tools and shop equipment will be found in the Appendix, as well as a list of tool suppliers.

2-12. A heating unit for compressed acetylene, manufactured by Prest-O-Lite, including a torch with a #1 tip, and a regulator and hose attached to a tank top.

METALWORKING PROCESSES

A few fundamental metalworking processes can be very practical in the preparation and finishing of metals used in textile constructions. These processes are annealing, soldering, pickling, wire drawing, beading, riveting, finishing, and polishing. They can all be accomplished with a minimum of shop equipment and work space. A small amount of time spent in practicing the various techniques and processes will result in sufficient skill and confidence to use them easily and effectively.

ANNEALING

Annealing is the process used most frequently in textile constructions because it relieves the stresses which build up in nonferrous metals as they become work hardened. This work hardening occurs very rapidly in sterling silver, brass, and 14-karat gold as it is bent, looped, hammered, twisted, drawn, or rolled through a mill. It occurs at a much slower rate in fine silver, 18-karat gold, and copper, and both pewter or 24-karat gold do not work harden at all. The stresses eventually cause the metal to fatigue and fracture unless they are eliminated by an even heating of the metal and a rapid cooling by quenching in water or a mild pickling solution. Any type of torch may be used for the heating process, but the flame should be large, fluffy, and tipped with yellow to ensure a slow and even heat.

Sheet metal must be heated gradually over the entire surface since most nonferrous metals are highly heat conductive, which means that any heat applied is instantly distributed throughout the sheet. Wire is best annealed in a tight coil (2-13), especially if it is very thin, in order to avoid melting or overheating.

The annealing process is more easily done in a semi-dark or dim environment since changes in color are the best indicators of temperature and these are not visible in bright light. Both natural and artificial light tend to obscure the color. Ferrous metals are annealed by being heated to a cherry-red color and allowed to air cool. Nonferrous metals are more properly annealed by being heated evenly until the entire piece is a dull red color and then cooled suddenly by quenching either in water or in a mild acid bath (pickling). Metals should never be picked up when they are red hot since they are liable to cracking at that stage. For that reason, they are allowed to cool in the air until all the redness is gone. At that point, the hot metal should be quenched by plunging it into water or a pickle bath.

2-13. Annealing a coil of wire. A #3 torch tip is used to produce a large fluffy flame. The wire is placed on a metal turntable covered with a soft asbestos pad.

PICKLING

Pickling is the process of cleaning metals after they have been heated. Metals containing copper will oxidize when heated, which results in a black discoloration. At times, this color change may be desirable, but if the original color is to be restored then the metal must be immersed in a pickling solution until clean. Pickle solutions can be made with noncorrosive and relatively safe commercial products such as Sparex or with a ten-to-one mixture of water and sulphuric acid. Both solutions are much more active when they are heated, and for this reason should be kept in a Pyrex or ceramic container which can be placed over an electric coil or a low gas flame. After removing from the pickle, the metal must be thoroughly rinsed with clear water and dried before being used. At no time should iron or steel tongs be used to remove nonferrous metals from the pickling solutions since this will cause another discoloration which is very difficult to remove. Instead, copper, bamboo, or heat-resistant plastic tongs should be used for this.

SOLDERING

Soldering is the process of joining two pieces of metal together with another metal of a lower melting point. Hard soldering is done at temperatures over 1000°F., and uses alloys of silver and gold which have distinctly different melting points. Soft solders, on the other hand, are alloys of tin or lead and melt at temperatures under 400°F. The soft solders must be used on low-temperature metals like pewter and lead, but may also be used on most of the other nonferrous metals as well (aluminum is an exception), although it is not as strong or permanent as the higher-temperature alloys. The hard solders, on the other hand, must be carefully selected by melting point in order not to melt the metal itself during the soldering operation.

The methods used for hard and soft soldering differ slightly but the basic procedures are the same and must be carefully followed to ensure good solder joints. First of all, the surfaces to be joined must be absolutely clean and perfectly fitted together. This is also true for wires which are to be joined. They must be thoroughly cleaned and fitted so that they touch each other along the entire joint. The solder itself must also be thoroughly cleaned before using and then cut into small segments for proper placement on the joint. Secondly, the solder joint must be well fluxed with a material appropriate to the type of solder to be used. In the case of hard soldering, this fluxing agent is a compound of borax and water which prevents the heat from oxidizing or soiling the metal where it is to be joined. To maintain the same clean metal, soft solders use a paste or a liquid flux, which is very different in character and composition from hard solder flux and is especially manufactured to suit a particular solder's alloy and melting point. These should be purchased for each specific solder since they are not readily interchangeable.

The application of heat is perhaps the most critical aspect of soldering and the one which causes the most difficulty for the beginner. All solders have a specific *melting* point that is not as significant as the *flow* point, which is the temperature the metal must reach for the solder to run and form a joint. Heating only the solder is not sufficient; all of the metals to be joined must be heated evenly until they reach the point at which the solder will flow. This heating should be accomplished as rapidly as possible in order to keep the oxides from building up and the flux from being consumed, since both factors severely inhibit the soldering process. The size of the flame is determined by the amount of metal being heated; the greater the quantity, the larger the flame must be to reach the appropriate soldering temperature. In addition, the flame must be directed to the area of greatest mass in order to maintain an even heat over the entire work to be joined. Solder flows toward the heat, and will flow indiscriminately onto whatever edge or surface reaches the proper temperature first. A joint will be formed only when all edges and surfaces involved are the same temperature *at the same time.*

After soldering, the flux is removed in boiling water, and the metal is restored to its original color by pickling. If the proper amount of solder has been used there will be a firm and clean joint; if too little has been used there may be gaps in the joint that will require additional solder and repetition of the entire process; if too much has been used the excess should be removed by scraping or filing.

WIRE DRAWING

The diameter of a wire can be made smaller by the process of drawing it through a series of graduated holes (2-14). The wire to be drawn must be in an annealed state at the beginning of the process and, if drawn many times, must be annealed periodically to remove the stresses caused by the compression and stretching of the material.

To begin, the wire is filed to a long (approximately three-quarters of an inch) taper and rubbed lightly with beeswax. The drawplate is placed horizontally in a vise with the numbered holes facing forward. The wire is inserted into the back of the drawplate in a hole one step smaller than its present diameter; the tapered end is grasped firmly with the draw tongs, and the wire is pulled smoothly straight through the hole. The process is repeated in successive holes until the desired diameter is reached, with the wire coiled and annealed after every four or five draws. The taper may also need refiling as the wire diminishes in size. The same process can be used to convert round wire into square, oval, half-round, or other cross-sectional configurations if suitable drawplates are available.

BEADING WIRE

Round beads can be formed on the ends of small diameter wires, providing a simple method of finishing multiple wire ends in a decorative manner. The wire is held with a pair of tweezers with the end of the wire to be beaded as far as possible from all surrounding metal. The torch is adjusted to a small, sharp, hot flame which is then applied to a small section at the tip of the wire (2-15). As the wire reaches the melting point the end fuses into a round bead that can be made larger by gradually moving the tip of the torch farther along the wire. There is a maximum size which can be supported by any given diameter of wire; beyond that, the bead will fall over to one side, and eventually drop off completely. Fluxing the end of the wire before heating produces a smoother flow. Copper wire does not bead easily because of its high melting point, and some brass alloys may be troublesome for the same reason. Fine and sterling silver, as well as all karats of gold, bead easily and smoothly if heated properly.

2-14. Drawing wire. A round-hole drawplate is placed horizontally in a vise. The draw tongs are used to pull the wire through a hole from back to front, thereby reducing the diameter of the wire.

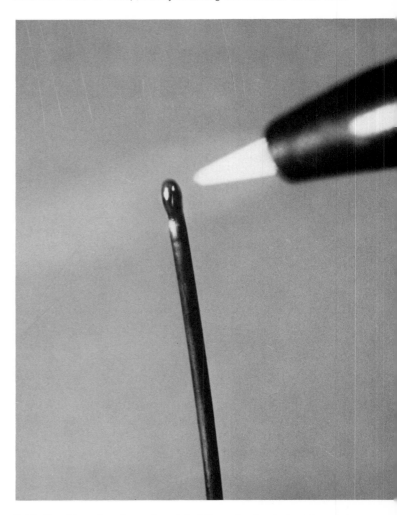

2-15. Beading wire. A small, pointed flame is directed at the end of a wire and as the wire melts a small bead is formed.

RIVETING

Rivets are connecting devices which require no soldering and are used to hold two or more pieces of metal together in a surface-to-surface relationship. They can be fashioned from brass or copper escutcheon pins (small nails with round heads), from tubing, from wire which has been beaded on one end (2-16a), or from a straight length of wire. Holes that are the exact diameter of the rivet or slightly smaller are drilled in the metals to be joined. Holes which are even a fraction too large produce very poor results. The nail or beaded wire is forced through the holes until the head or bead is flush with one surface (2-16b). The remaining wire is clipped slightly above the hole (2-16c), and the rivet is set by stretching the wire over the hole with a wedge-shaped hammer while resting the rivet upright against a block of steel (2-16d and e). When using straight wire, the clipping and stretching process is followed on both ends of the wire. Tubing must be cut with a saw and is more easily set by flaring the tube over the hole with a tapered punch hammered into the opening. Rivets of all kinds may be filed flush with the surface to form an inconspicuous connection, or they may be left intact with the heads forming a decorative as well as a functional element.

2-16a. Making a rivet. A small piece of wire with a beaded end is placed in a close-fitting hole in the drawplate. The bead is flattened with the round end of a small rivet hammer.

2-16b. Setting a rivet, step one. The rivet is pushed through tight-fitting holes in each of the layers to be fastened together.

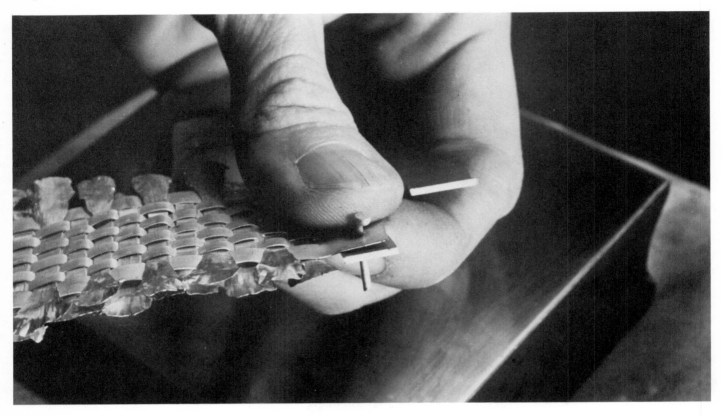

2-16c. Setting a rivet, step two. The unfinished end of the wire is clipped close to the last hole.

2-16d. Setting a rivet, step three. The unfinished end of the rivet is stretched by hammering with the wedge-faced end of a small rivet hammer while resting the opposite end of the rivet on a steel block.

2-16e. Setting a rivet, step four. The rivet is completed by hammering the stretched end with a smooth, flat hammer head.

FINISHING AND POLISHING

The intricacy of most textile surfaces and structures precludes the use of standard abrasives and buffing compounds on completed work. Any strenuous treatment of the surface of either sheet metal or wire to remove dents and scratches must be applied to the various structural elements before they are assembled. The same is true if highly polished surfaces are desired on all or part of the piece since the various inter-linkings, interlacings, and interlockings impede the polishing operation and, in fact, could themselves be damaged by machine buffing and polishing.

Strips or large sections of sheet metal should be completely finished and polished in advance. This may be accomplished in a variety of ways using a combination of abrasives and polishing techniques, but should be preceded by any annealing operations that might be necessary. The discoloration caused by heating the metal is easily removed along with scratches and other unwanted surface blemishes during the finishing procedures.

2-17. Hand finishing tools. *From left to right:* three grades of Scotch stone, a hematite burnisher, a steel burnisher with small scraper on the other end, and a three-sided hollow scraper.

The surface is rubbed vigorously with a medium grit emery paper or cloth (#240, 300 or 320) to remove surface scratches and minor imperfections. Very deep scratches are removed with a Scotch stone (2-17) and water, or with a hand scraper (2-17) that "removes" the scratch by lowering the surface of the metal in the area of the scratch. The surface is rubbed in one direction with finer and finer abrasive papers until an even patina of delicate surface scratches produces a matte finish. If a reflective surface is desired, the metal may be buffed vigorously with a felt stick heavily coated with an abrasive compound like White Diamond, or the same compound can be used on a felt or coarse muslin buff on a low-speed polishing machine. A highly polished surface may be achieved by rubbing vigorously with a chamois stick impregnated with jewelers' rouge or by using rouge on a soft muslin or chamois buff on a low-speed polishing machine. All abrasives and polishing compound are then removed from the metal by scrubbing thoroughly with a soft brush in a hot solution of soap, water, and a few drops of ammonia. The metal is then rinsed in hot water and thoroughly dried to prevent water spotting.

A highly polished surface is not the only kind of finish possible, nor is it necessarily the preferred one. Matte finishes on sheet metal can be achieved easily with a simple sequence of emery papers from medium to very fine (#600) or at any stage in between. Fine steel wool can be used on highly textured surfaces to bring up a soft polish or it can be used on smooth surfaces after the emery papers to produce a finer finish. Fine steel wool saturated in liquid detergent is an excellent agent for a final polishing after a piece is completed, even when highly polished elements have been used in the construction. It can be followed by a highlighting on raised areas using a rouge-coated chamois stick held in the hand. It is important to note that under no circumstances should a *finished* piece be polished by any kind of machine.

On wire fabrications, much less in the way of finishing or polishing is required, provided the material is clean to begin with. Working with coated copper wire, of course, eliminates all need for polishing since the color and polish are permanent, provided the metal is not subjected to abrasives, heat, or acid treatment. The natural finish of fine and sterling silver, gold, brass, and uncoated copper wires will be slightly dulled as the result of annealing and pickling. The wire can be rubbed lightly with fine steel wool to restore the original color and luster before it is used, although that is not absolutely necessary. The total structure can be treated in the same manner when it is finished, and the final effect will be equally satisfactory.

A simple and highly effective finish can be achieved on all types of textile surfaces after they are completed, whether they use wire, sheet, or a combination of the two, provided no coated copper is used in the structure. A hand brush of fine brass wire, saturated with a soapy lubricant such as liquid detergent or green soap, may be scrubbed over the entire surface to bring up a beautifully soft polish. This is especially effective with fine and sterling silver, both of which turn white after annealing and pickling. The white surface is changed to a brilliant but soft luster when "scratch-brushed" with brass-wire brushes, which are better able to reach down into the recesses than pads of steel wool. Such brass-wire brushes are also available for use on low-speed rotary polishing machines that might be used on a metal textile structure of fairly rigid form and closely worked surface. All other structures should be hand-finished only since they are easily caught in the machinery, a situation which is both dangerous to the finisher and disastrous to the work.

ELECTROPLATING

Electroplating is the process by which thin layers of gold, silver, nickel, copper, chromium, etc. can be deposited evenly over the surface of other metals. It is done after a piece is totally finished, including cleaning and polishing, and is used as a means of changing its color, although with very delicate textile constructions the plating may also add body to the piece.

Through the use of lacquer stop-outs, the plating can be done selectively in some areas and not in others, allowing for a two- or three-color effect. Given the fact that a great variety of metals can be used to produce strong color effects, or the same metal can be used with different patinas and alloys to produce delicate and subtle color modulations, selective plating can produce an even greater range of effects. For example, if a piece fabricated in copper has some areas silver-plated and some areas gold-plated, it can have all three colors present simultaneously without the expense of working in silver and gold directly. Since the plated metal skin is very thin, it will not obscure scratches or blemishes. These must be removed first either by hand, emery papers, Scotch stone, steel wool, or burnishers or by machines, using buffing wheels and compounds. For plating purposes the metal must be chemically cleaned and absolutely free of oil and grease. This thorough chemical cleaning can be done most effectively by the plater with solutions and processes especially designed for that purpose. Unless proper equipment and knowledgeable technical assistance are available, it is recommended that the services of a commercial plating firm be used for this process.

ENDINGS

In all of the textile processes, endings need to be given careful consideration to ensure permanency as well as refinement of the finished form, whether it be a small piece of jewelry or a large wall hanging.

There are two quite distinct approaches possible, one utilizing standard metalworking processes such as soldering and riveting and the other remaining closer to textile methods, with the use of twists, coils, braids, and other improvisations.

If one has access to metalworking tools and is knowledgeable in the use of basic techniques, then it is simple to build separate clasps and fittings out of sheet metal or forged wire. Fittings can also be fashioned in wax and cast in metal, although care must be taken when using this approach to keep the weight as light as possible. Heavy clasps and fittings can easily cause distortions in the form of pieces made from lightweight wire. Another very simple device for ending wires is to bead the ends of single or multiple elements by melting them quickly with a small torch as previously described in the section on beading. This is particularly easy and effective with gold and silver wires because they bead very readily and can be hammered afterward to form small flat disc endings that catch the light with great brilliance.

Rivets are "cold" connections which do not require the use of complex metalworking tools, and can be made easily from small brass or copper nails or from small-scale tubing of any soft metal. As previously described, holes must be punched or drilled in a size close to the diameter of the rivet to ensure a proper fit, the nails or tubes must be placed in the holes, clipped or cut close to the surface, and hammered lightly on a steel block to spread the ends on both sides so they can no longer slip through the holes. Commercial two-part rivets for leather or fabric can also be used, although they are sometimes difficult to find in small sizes.

The other approach, using combinations of textile methods, requires no tools or equipment beyond those already collected for the various processes themselves. For example, a simple crocheted chain stitch makes an excellent finish for a knitted piece and can be expanded to become an important decorative element as well. Ends can be worked back into the structure of both woven and braided fabrics that are dense enough to absorb the extra thickness without noticeable distortion. Longer ends can be left hanging to be finished later into braided edges, or to be made into separate small coils by winding around the back end of a small drill bit or a nail. The coils can also serve as holders for inserted feathers or other decorative elements such as beads and pearls. Wires can simply be twisted, folded, and twisted again to form dangling ends, or can be twisted with other materials for varied effects.

Endings of any sort should be compatible with both the form and the movement of the piece. The ending, however inconspicuous or lavish, requires careful selection as an important part of the design. Different kinds of endings can change the whole character of a piece, and several improvisations should be tried, evaluated with sensitivity, and replaced if necessary with new solutions, before the final construction.

3/Weaving

Weaving is the interlacing of two separate sets of elements to produce a fabric. The element called the warp is set down first, usually in a parallel arrangement, although in some systems such as pin weaving it can be radial or even asymmetrical. The element which interlocks with the warp is called the weft, and creates a stable planar structure from the two sets of originally linear elements. Usually it is perpendicular to the warp.

Weaving allows a seemingly limitless range of variations within a basic structural system and is ideally suited for use in metal because it does not require too much flexibility in the elements, and the structure also benefits from their rigid support. Metal weaving can produce a wide range of effects with and without the usual structure systems, from jewelry and small-scale objects in precious metals to architectural-scale wall reliefs (see 9-1) or free-standing sculptural forms in weather-resistant metals, which have the advantage of being able to survive outdoor conditions in which yarns would be totally impractical.

It is possible to weave such small-scale objects as one-quarter inch bands for rings (3-1), three-dimensional cubes and cylinders for use as beads (3-2), and narrow strips for bracelets, neckbands, and ornamental devices on other structures. This scale in metal weaving has its historical precedents in the incorporation of flat woven metal strips into lace fabrics to form small dimensional nodules, which sometimes stand a half inch above the background and have the effect of jeweled encrustation. On the same construction principles, larger scale bracelets, belts, collars (3-3, 3-3a), hats, and even simple articles of decorative clothing can be made. The only adjustments which need be made are in the choice of materials—the gauge should always be appropriate to the scale—and in the manner of supporting the structure while it is in progress. The only limitations are those which the artist imposes for aesthetic or physical reasons.

Weaving can also be accomplished in the round, with or without structural devices, and in this area of three-dimensions the use of metal is superior to more usual weaving materials, since it can achieve stability and strength without the necessity of structural supports. The rigidity of the metal is a very definite asset where the flexibility of yarn would have been a disadvantage (3-4).

3-1. Ring by Steven Brixner with woven gold decorative element on silver and ivory inlay. The stone is green serpentine. The woven element is 14-karat gold with a weft of wire and a warp of 30-gauge sheet cut into ¹⁄₁₆-inch strips.

3-2. This small one-inch copper cube was woven in the hands without the use of board or frame. Strips of the same size sheet metal are used for both the warp and weft.

3-3. Woven neckpiece by Mary Lee Hu. Both the warp and weft elements are braided wire in fine silver, fine gold, and coated nickel alloy. The weaving is done with an eccentric warp which becomes the weft in the front section where the two sides merge in a plaid pattern. A heavy outer structure of sterling silver helps to maintain the form and provides a strong visual outline. The total length of the neckpiece is 24 inches.

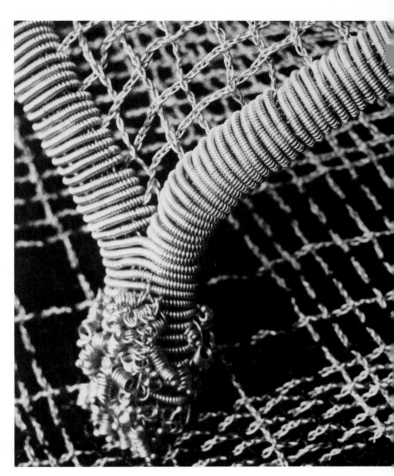

3-3a. A detail that shows the braided wires functioning as warp and weft. Notice the continuous coiled wire, which is used to bind the edges of the form, and the small coils that provide a decorative flourish to the finial in the front of the neckpiece.

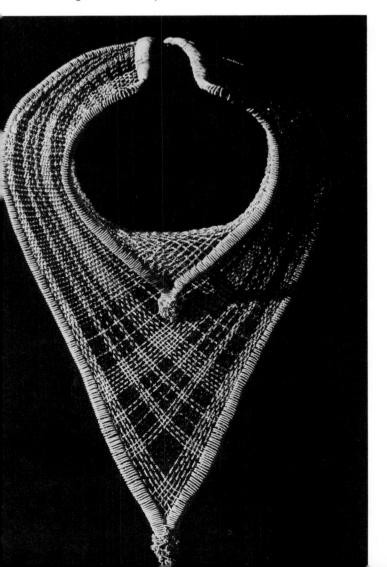

3-4. A series of small, 1- to 2-inch woven baskets made in Germany and intended as table favors. Combinations of wire and sheet metal strips are used to produce different shapes and textures in the round.

STRIP WEAVING

The weaving of metal can be accomplished in many ways, depending upon scale, type of metal used, and the desired end products. The simplest method of weaving strips of metal off the loom is to fasten the warp with masking tape or nails, at one end only, to a frame or board. The warp can be created from a single sheet of metal that retains its original shape by not having the warp cut all the way through (see 3-9a). This allows the construction to be accomplished with a minimum of trauma to the metal and with a maximum flexibility in the adjustment of tension and development of textures. Thin sheet metals and wires of all sorts adapt well to the weaving process unless, as already mentioned, they are too soft to retain the necessary structure unaided. Weaving does not violate the inherent qualities of metal, as the elements are not required to stretch or convolute in ways which are difficult or impossible for the material. They need only go over and under each other, creating intriguing surface textures at the same time they are building structures of far greater flexibility than is characteristic of the original material.

There are vast textural changes which can result from simple changes in warp and weft composition and in variations of weave constructions. Together these produce a wide range of effects. The most commonly used weaves are plain weave, or tabby (one over, one under; see 3-5, 3-6); basket weave (two over, two under; see 3-8); and twill weave (usually one over, three under in a diagonal rotation; see 3-8). Weave constructions can be much more complex than these examples, and the weaving books listed in the Bibliography will provide further details.

A wire warp gives quite a different surface quality to a woven structure. A wire warp with a weft that combines wire and strips of metal can produce many varied effects. Slight structural changes such as these will influence the way the piece moves and flexes as well as its appearance.

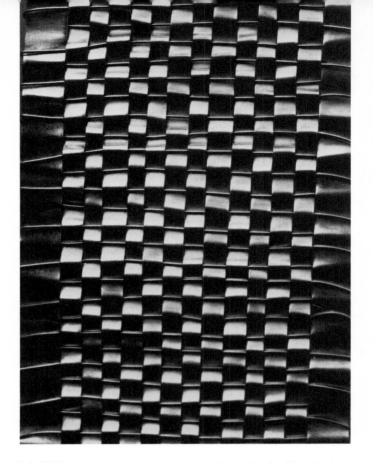

3-5. Plain-weave construction in fine-silver wire (weft) and sheet (warp).

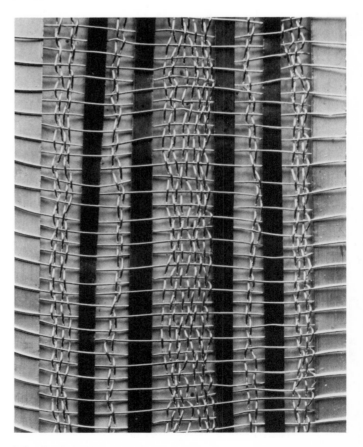

3-6. Another fine-silver plain weave with warp of alternating wire and sheet and weft of wire.

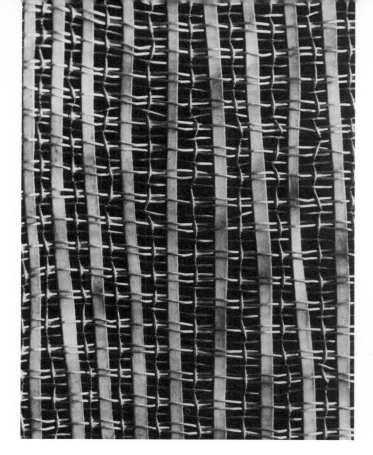

3-7. In a variation of plain weave, the tabby shots alternate with shots of one over, three under. Warp alternates round and flat wire of fine silver.

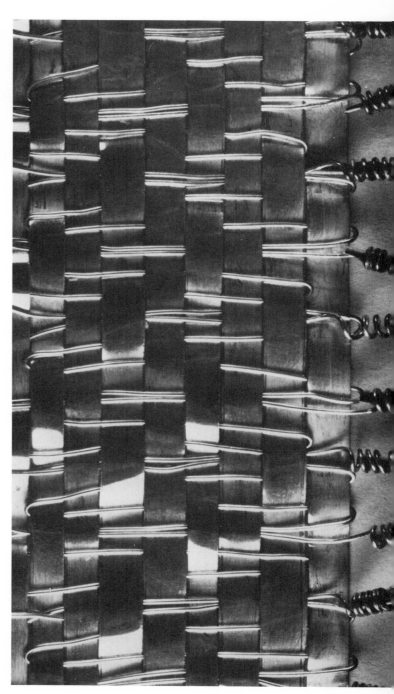

3-8. Basket weave in a twill rotation, which produces a diagonal pattern. The warp is fine-silver sheet and the weft is fine-silver wire, yet the effect is very different from 3-5, in which the same elements interlace in a different weave construction.

Weaving strips without a loom is the approach I would recommend at the outset, since it offers the greatest area of experimentation with a minimum of equipment and prior knowledge of technical procedures. You will need a soft bulletin board or cork board, large enough to hold the entire piece of weaving, nails or pins easily pushed into the board, masking tape, and a pair of large scissors.

The materials can vary but I suggest copper or brass foil, brass shim stock, thin sterling or fine silver, or thin aluminum sheet. They should be approximately 30 to 36 gauge and preferably in soft or annealed state, although this is not absolutely essential.

The procedure is quite simple. First construct the warp by making parallel cuts in a single sheet of metal or by placing separate warp strips side-by-side to produce the desired width. Fasten the total warp to the board at one end only, with a nail or pin through each warp element. Masking tape can be used on the top to hold the warp ends firmly (3-9a; 3-10a).

To weave, lift the appropriate warp ends for each weft shot with your fingers and pull the warp through. A great many effects can be achieved by varying the metal used for weft (different forms such as wire and sheet, different color of metals, different shape of strips), and by different weave constructions. Carry the weft across and insert pins on either side to control the widths of the finished piece. Push the weft as tightly as desired before lowering the warp ends.

Continue with the weaving, placing masking tape across the whole structure periodically to secure it firmly during the weaving process. Only the last weft shot needs pins to hold it in place. The other pins can be removed, since the new weft shot always holds the previous one in place (3-9c, 3-10c). Depending on the material used for warp and weft, one element may be more flexible than the other. In 3-5, the weft lies straight without distortion, while the warp has been moved up and down, over and under, the weft shots, creating quite different light-reflective qualities from a completely flat surface or from the more weft-faced surface that might be achieved by distorting the weft and keeping the warp flat (3-9d, 3-10d, 3-11, 3-12).

When the piece has been finished, both warp and weft must be fastened in some way. The warp can be fastened by folding over, weaving in, riveting, or soldering each end to the one next to it or to a separate piece (depending on the way the strip will finally be employed). If one end of the warp has been left uncut (see 3-9c, 3-10e), only the loose end will have to be secured.

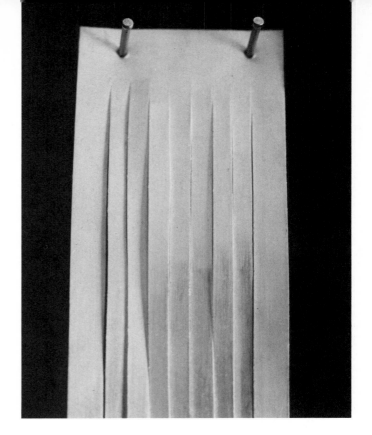

3-9a. The warp. The warp strip is a single sheet of 30-gauge sterling silver, 1⅜ inch in width. Each warp element is ⅛ inch wide except for the selvages, which are ¼ inch wide. Nailed to a soft board only at the top, the strip was cut with ordinary scissors after being measured carefully and scribed.

3-10a. A warp strip in a single sheet of brass shim stock, which has been annealed and cut free hand with scissors into five strips of a wavelike shape. No metal has been removed from the initial strip except for the outside edges. The surface has been textured by stamping.

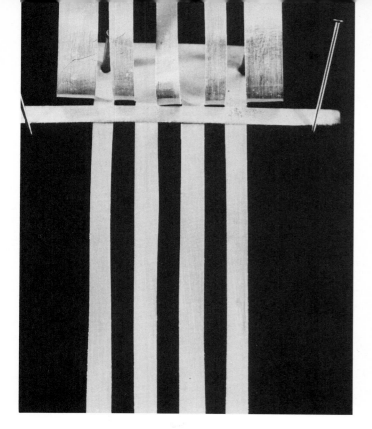

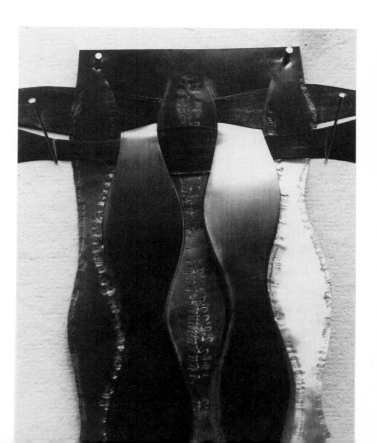

3-9b. The first weft shot. The weft is 30-gauge sterling silver and is cut in strips ⅛ inch wide. The weaving is accomplished by actually picking up every other warp strip and bending it toward the top of the board. The weft shot is held in place by the insertion of two pins, one on either side of the warp.

3-9c. The second weft shot. The warp ends for the first shot have been lowered and the alternate ends raised.

3-10b. The first weft shot. The weft is sheet copper, cut freehand with a scissors and repeating the wavelike shapes used in the warp. As in the last series, the weft has been put in place by lifting alternate warp ends, inserting the weft strip, and pushing it as far toward the top as possible.

3-10c. The second weft shot fastened with a straight pin on each side of the warp.

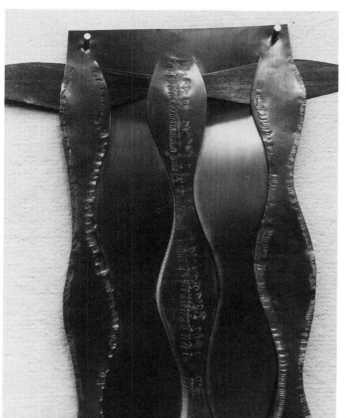

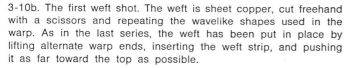

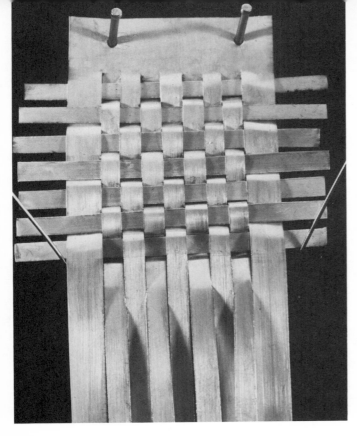

3-9d. Seven weft shots. The structure of the plain weave can be clearly seen, with a flat weft and slightly dimensional warp.

3-9e. Completed woven strip showing all weft ends woven back into the structure, after which they are tacked in place with solder.

3-10d. Completed woven strip. The structure of the plain weave looks very different here because of the undulating outlines of warp and weft, and the visual emphasis on linear rhythm rather than surface texture.

3-10e. All warp and weft ends have been turned to the back of the strip and fastened with rivets.

The weft also must be secured. If using a continuous wire for weft, only the beginning and end of the wire must be woven, folded, or soldered (3-11, 3-12). Otherwise, all weft ends must be secured so there are no loose ends at the edges of the finished piece. In 3-9c the weft has been fastened mechanically through the use of silver solder on the back, and the entire strip is flexible in the direction of the warp. In 3-10e, all warp and weft ends have been turned to the back of the strip and fastened in place with the use of silver rivets. This mechanical means of fastening makes it possible to produce woven forms without benefit of soldering or welding.

3-11. A warp arrangement in which sheet-metal strips and individual wires are used as alternating warp elements and in which the weft is continuous wire. Note that it is held in place at the proper width by pins at the beginning and end of each weft shot.

3-12. A bracelet sample, woven in a nonrectangular shape, uses as warp a closely set series of very thin strips of copper and pewter alternating with wire elements. The weft. is a continuous wire which is pulled tightly, forcing the sheet-metal warp ends to move dramatically over and under the weft. By pulling the weft gradually tighter and then looser as weaving progresses, the center of the piece is contracted. (Courtesy of the Crafts Council of Australia.)

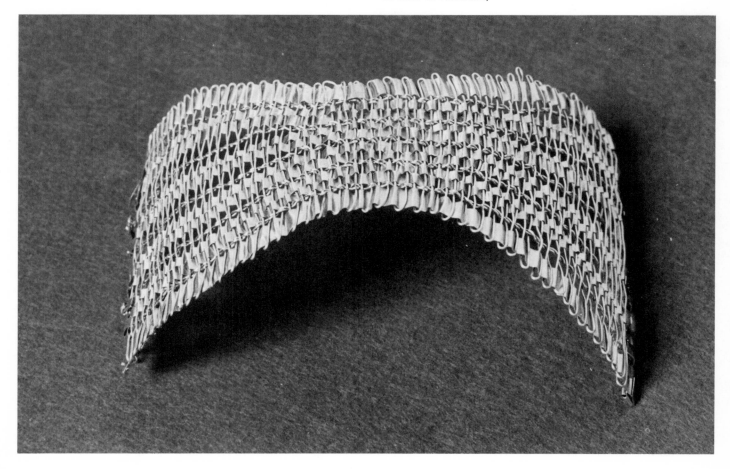

3-13. Necklace constructed of a twill-woven band in sterling-silver sheet and fine-silver wire. The feathers are held by supplementary wires inserted into the structure and coiled at each end.

SHAPED WEAVING

It is not necessary for weaving to be done only in variations of the band or rectangular format. When working without a loom it is possible to weave to any shape by controlling the edges of both warp and weft. Both flat and silhouetted forms can be shaped through structural manipulation and placement of a wire warp in a shaped outline, such as the one-inch open-weave heart (3-14) which was woven around a series of nails placed in the outline of a heart.

The technique for making such forms is called pin weaving, and it is done on a board with nails or pins in it. The warp and weft are wrapped around them, and, after the piece is completed, slipped off them again to free the weaving from the background structure. No additional support is necessary when metal is used. The usual procedure is to wrap a parallel warp around the pin outline, but circular and semicircular forms also can be created by using a radial or other nonparallel warp. The weft is then woven across the warp and around the pins, so that it conforms to the same outline.

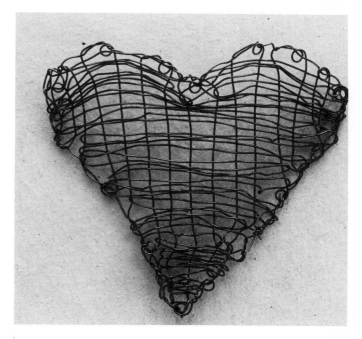

3-14. Pin-woven heart form in 30-gauge copper wire.

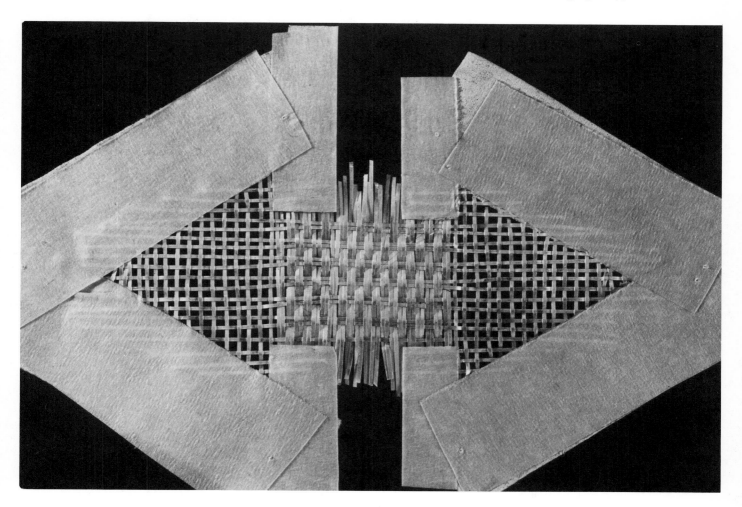

3-15. In a more easily manipulated variation of pin weaving, warp ends are placed between two strips of masking tape, enabling the warp to flex sufficiently to receive the weft shots without distortion. Angular forms such as the arrow are easiest to achieve when using this technique in metal.

Instead of pins or nails, masking tape can be used when working with metal. The advantage to this procedure is that it allows the whole taped warp to be picked up and flexed to permit the easy insertion of the weft. If pins are used with a nonstretchable material such as metal, it is very difficult to lace the weft over and under the warp, which is rigid and flush against the background. The possibilities for structural variations are the same as in strip weaving, but the finished shape is not confined to rectangles, although angular forms are the easiest to achieve. The arrow (3-15) was made this way and incorporates both plain weave and basket weave in a varied surface texture and density. The ends of the sample could be finished either by soldering to a frame around the form or by being individually bent to the back and woven into the structure.

The substitution of a dimensional form for a flat board enables the weaver to produce self-supporting volumetric forms with the basic techniques of pin weaving. Warps can be nonparallel in many other ways than radial—they can zigzag, intertwine, split and reunite, or remain open and expanding in multiple sections. The only requirement is that they produce a stable structure with the weft. No armature is necessary when metal is used, but with very malleable materials such as thin silver wire, the form can be created over a temporary support. Finishing cannot be achieved by weaving the warp ends back into the structure if the weave is too open to hide them. In such cases, wrapping and coiling might be good.

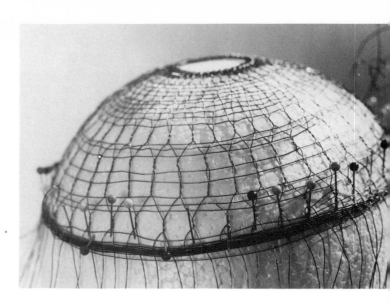

3-17a. A hat is woven to shape directly on a Styrofoam (plastic foam) form using a leno weave on a radial warp.

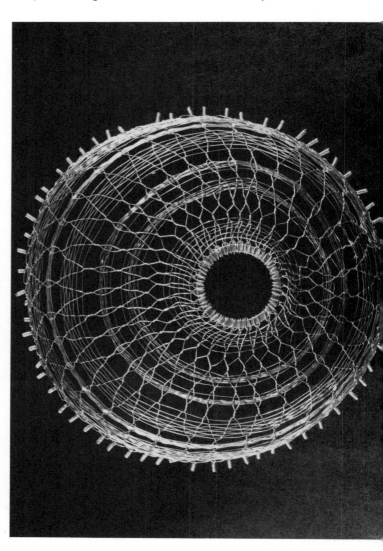

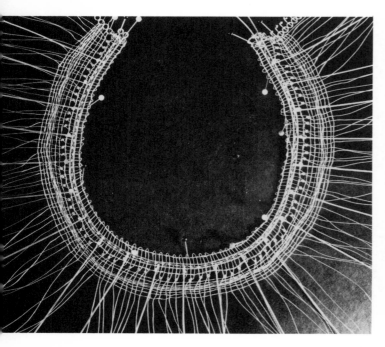

3-16. A partially completed woven collar in fine-silver wire of 26 and 30 gauge, with additional warp elements of forged sterling-silver wire. The warp is arranged in a radial pattern with the warp ends looped into a separately woven narrow band at the top.

3-17b. After weaving, the hat is removed from the background form. Notice the manner in which the upper warp ends were knotted over a forged ring in the center before weaving began.

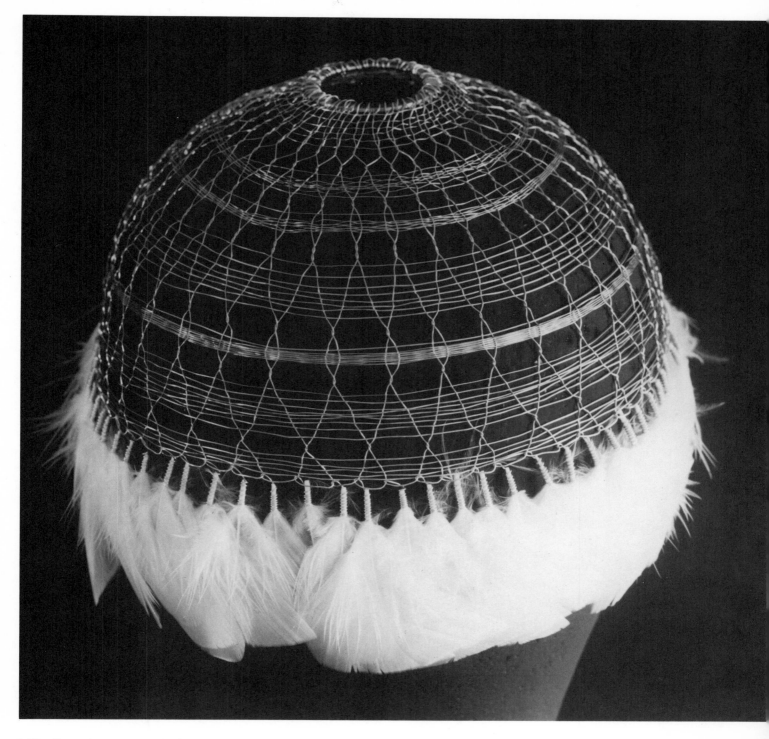

3-17c. For a decorative finish, white feathers have been inserted
into warp ends tightly coiled over a small drill bit.

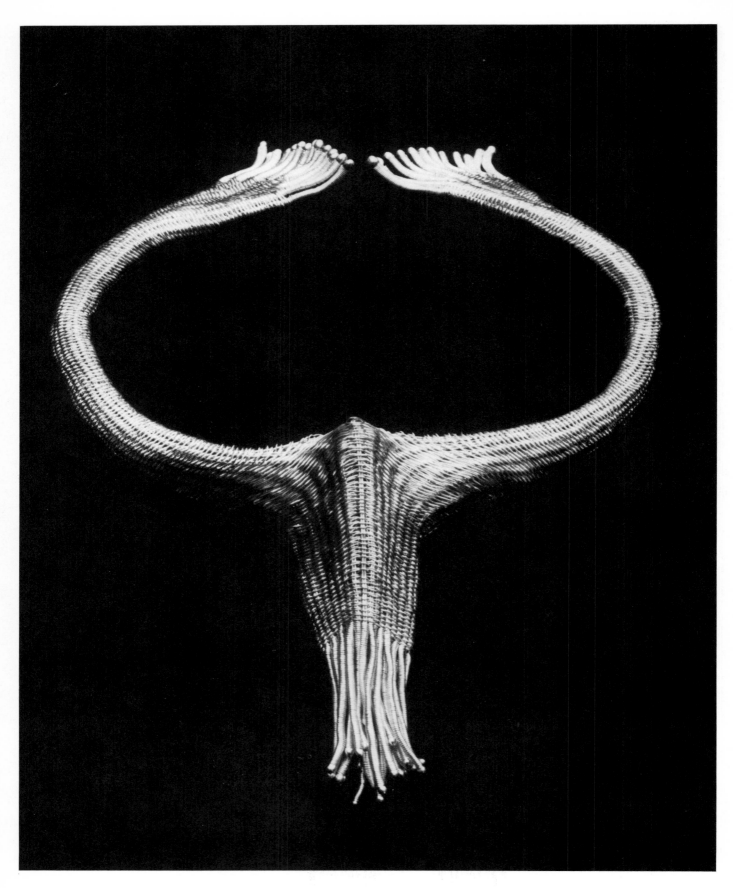

3-18. A woven neckpiece by Mary Lee Hu. The weaving is done in fine-silver wire over a rigid structure of sterling silver to produce a torque-like form. Tightly coiled wire is wrapped around the warp ends to give a decorative finish to the piece.

CARDWEAVING

This technique is used to produce narrow strips of patterned weaving where the weft is completely hidden (a warp-faced fabric) and the warp threads of different colors make a pattern of stripes, arrows, diagonals, bars, or other simple figures by means of a twisting motion of the cards through which the warp is threaded. This twisting warp is a unique characteristic of cardweaving.

The pattern is planned beforehand on graph paper with light and dark squares indicating the threading of the different warp colors on the cards. When using metal, annealed copper or brass wire in 26 to 28 gauge is a good choice because it can be obtained in a range of colors and shades. As mentioned, coated wire comes in contrasting colors, and can be used without fear of the coating wearing off, as long as there is no heat or chemical treatment applied during the finishing of the piece and it is not subject to abrasion in use.

In addition to the wire, you will need square cards made of cardboard, plastic, or thin Masonite. Ten cards will create a warp of about one half inch in width, which is suitable for a headband, thin belt, or other decorative strip. The cards should be cut to equal size squares of three or four inches, and have four holes punched in them (with a hole punch or drill), one at each corner, but not too close to the edge. These holes must be lettered ABCD, in clockwise sequence from the upper left corner, and each card should be numbered clearly for position.

Next, the design must be planned on graph paper, as shown in the diagram. This particular pattern is for an approximately two-inch band (40 cards) and shows eleven weft shots—only eight shots are really necessary to plan because the pattern repeats after that. Note that the color of the weft is not indicated on the diagram in any way; this is because it will not show in the warp-faced fabric. Even the weight of the weft will make very little difference.

The diagram is used to help thread the cards. Since 40 cards are used, a total of 160 (40 x 4) warp ends will be needed. If one counts all the black squares in the first four vertical columns the total is 88—therefore 88 warp ends of one color will be needed, and should be cut to the required length of the finished piece plus 24 inches for take up and threading. Again, count the white squares in the first four columns, and the total indicates that 72 ends of the other color must be cut. The number of dark plus light wires should of course equal the total, which is 160 in this example.

Now the cards are arranged in numerical order, each with the A hole in the upper left-hand corner. To thread the first card in this example, use a dark wire for all of the holes. Continue in the same way for the next three cards, but for card number 5 use a light wire in holes A and D, etc. Make sure that all the wires are threaded through in the same direction,

from the back to front (or front to back) of the card. The end of the warp that has been threaded through is pulled a short distance, about 6 inches, and then can be held temporarily under a book or brick. After each card is threaded, it is turned face down and each succeeding card is placed on top of the former ones in the same position, that is, with all the A holes together. When all the cards are threaded, they are tied firmly in a packet using string or wire, and the end of the warp that has been pulled through is taped firmly in a flat packet after it has been spread to the desired width of the band.

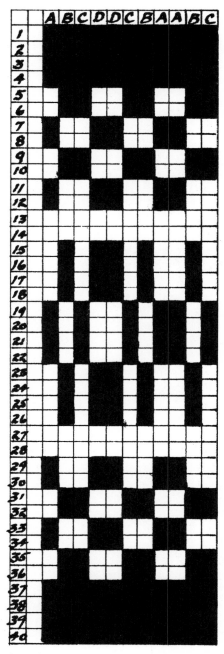

3-19. Cardweaving diagram. This is a graph for a 40-card warp in a light and a dark color. Eleven weft shots are shown, and the letters indicate that the cards are to be turned four quarter turns forward and four quarter turns backward. The sequence repeats after the eighth turn, although three extra turns are shown.

Setting up the warp to weave requires a method of providing tension on the warp wires. In some systems of cardweaving, the warp is tied and knotted over a hook at one end while being attached to a backstrap at the other. When working with wire, a good method is to tape each end packet of warp to a block raised above a board or table that is the length of the warp. The blocks have to be high enough to allow the cards to turn as weaving proceeds. Each end packet of warp is taped securely to the top surface of its board, and then, just as on a warp beam, is rotated ninety degrees forward to increase the tension, and taped securely again.

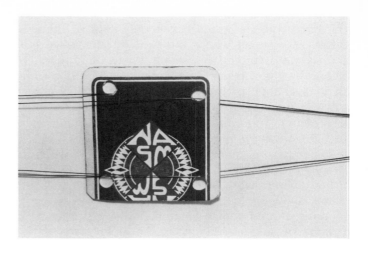

3-20a. A single card is threaded for cardweaving. All four warp ends are threaded in the same direction, from front to back (shown in photo as passing in front of the left side and behind the right side of the card).

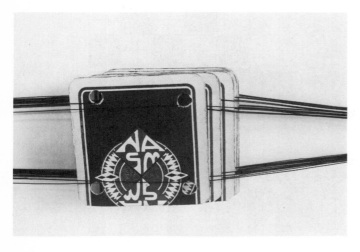

3-20b. The cards are stacked on top of each other with all the A holes in the upper left-hand corner position. The wires form an opening, called the shed, which receives the weft.

3-20c. The cards are rotated one-quarter turn at a time to change the shed, and the weft is inserted after each turn.

3-20d. One end of an improvised frame which holds the cardweaving under the necessary tension. Masking tape is used to secure the warp ends to a block of wood which is then tightly clamped to a frame or table.

Before weaving, the packet of cards is untied. The tension of the warp should keep the cards suspended and in order with the wires automatically divided into two layers that form an open space called the shed. To space the warp evenly, four strips of metal or cardboard, about a half-inch wide, are woven in before the first weft shot (see right-hand side of 3-20d. The strip is inserted into the opening of the shed and the cards are turned forward one-quarter turn so that the B hole is now on top (3-20b).This is repeated three more times, and each time the strip is beaten down firmly to help space the warp, until the cards are again in their upright position with the A hole in the upper left-hand corner.

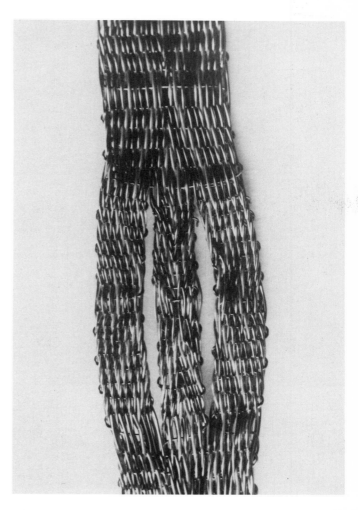

3-22. Detail of a cardwoven band of 26-gauge copper and brass wire separated into three sections by the same method. Woven by Linda C. Bell.

3-21. A cardwoven band of 26-gauge copper and brass wires with slits in it, made by using two separate wefts and rotating each group of cards separately. Woven by Linda C. Bell.

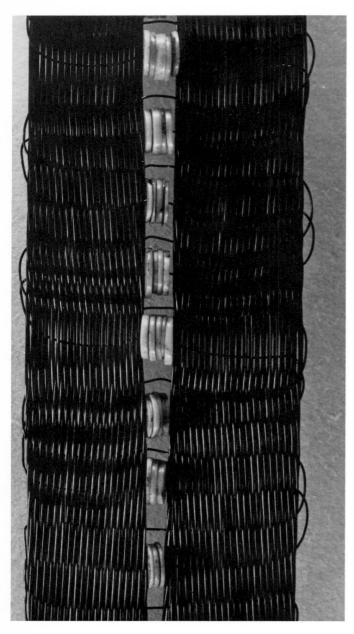

3-23. Beads inserted into the center of a cardwoven band of 28-gauge copper wire in two colors. In order to achieve this effect, beads are strung on the weft and then pushed into place as necessary when the weft is inserted in the shed.

By the time the four strips are in place, the warp is evenly spaced across the width of the strip, and the beginning of the weft element can be inserted. It too must be beaten down firmly into the shed with a thin stick or metal ruler. The weaving continues with the weft wire being placed into the shed at every quarter turn of the cards and beaten into place both before and after the next quarter turn. This ensures a firm and even weave. The cards are turned in the preplanned order of the diagram (see 3-19), which in this case indicates four forward quarter turns (ABCD) and then four backward quarter turns (DCBA). Many other variations can be used, such as moving the cards forward all the time or forward eight/back four/ forward two/back two. Each sequence of motions will alter the pattern, but it should be noted that unless the forward motions are equalled by backward motions, the warp will continue to twist and eventually inhibit weaving. To avoid this, especially when the cards are being moved forward all the time, the warp ends on the side of the cards away from the weaving may be periodically released, untwisted, and retaped.

Weave variations are endless, not only in the graphing of patterns, but in the manner in which the cards are used after they are threaded. For example, patterning is altered by different sequences of card turnings, as mentioned earlier. Cards also may be divided into separate groups which move in varying sequences; for example, ten cards on each side can move in regular sequence while 20 cards in the center move forward only. This will create pattern variations within the woven structure. Warp threads can be crossed by changing the position of one or more cards at any point during the weaving process. For example, cards number 5 and 6 might be inserted between cards number 1 and 2, producing a cross in the weft at the point in the band where they were moved and altering the color pattern.

The warp may be woven in different sections to create slits (3-21 and 3-22). To do this, separate wefts are used for each warp section, and the cards that control each section are turned as a group. Another effect can be created by stringing beads on the weft as it is threaded into each shed, either within the woven strip (3-23) or on the edges.

A form of double weave may be achieved by using one weft on the upper layer of warp threads and one weft on the lower layer. Variations in the width of the band can be created by pulling the warp threads open after the weft is inserted, or by gradually spreading the warp ends throughout a passage of weaving.

LOOM WEAVING

Metal can be woven on many other looms besides those for cardweaving and pin weaving. Frequently it has been stretched across fixed frames and hoops in the creation of screens and hangings, and used as weft where rigidity of structure is desired in a woven piece. With the proper selection of wire type and size, wire warps, too, can be used as easily as yarn warps on inkle looms, backstrap looms, and even standard two-, four-, and eight-harness looms. It can be threaded with as much ease and variation, and different sizes and colors of wires can be woven in all manner of weaving drafts and densities of weaves. There is virtually no limit to the weave patterns and surfaces which can be created with wire on a loom, from the simple plain and twill weaves already described in the section on strip weaving to the most elaborate overshot and open weaves. In fact, wire is ideally suited to open weaves (3-24) because the twist of warp or weft which locks such weave structures in place is much firmer and more permanent than it would be in yarn.

The information that follows refers to the special methods or precautions that should be used when weaving with wire; for basic instruction on weaving itself, see the books in the Bibliography. The series of samples illustrated were all woven on an eight-harness Structo loom with a warp of 32-gauge copper wire. This material is malleable enough to be threaded through the harnesses and wound around the warp beams, but is also strong enough not to break with the tension put on the warp. Other materials that are also suitable for weaving are gold and fine silver wire of thin gauges.

A few necessary precautions should be taken when preparing a metal warp for use on a loom. In winding the warp it is preferable to do so around as long a length as possible in order to avoid crimping the wire by unnecessary bends. Pegs in the ground or nails in the wall will serve adequately as a warping mechanism. Before removing the warp from tension on the pegs, place masking tape across the width at frequent intervals to eliminate unnecessary tangling. The relative inflexibility of metal makes chaining the warp impractical. The warp should be kept taped until one end is fastened to the back beam of the loom, and spread to the approximate width of the warp.

When winding the warp around the back beam, it is necessary to use both paper and slats to separate the layers of warp as they wind around. Constant checking is necessary to keep warp ends uncrossed and in as even tension as possible.

After threading through the heddles and reed, the warp is attached to the front beam (not by tying as with yarn but by twisting each piece around itself) in small sections with careful adjustment of the tension of each warp end. The maintenance of tension is not a serious problem at all despite the lack of stretch quality in wire. An even tension must be established across the width of the warp, and if the warp will be subjected to extremes of movement, it should be loosened slightly first.

3-24. A grouped-warp leno weave sometimes called Brooks Bouquet. This sort of open weave is very suitable for metal, which locks it securely in place. All loom-woven samples are by Steven Brixner.

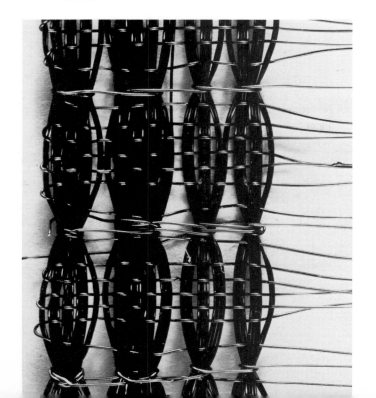

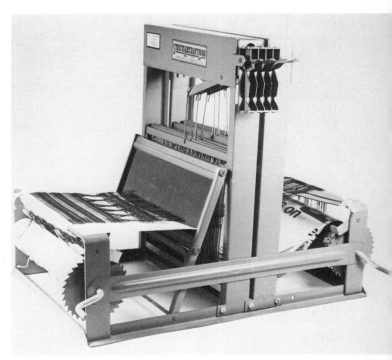

3-25a. A table model eight-harness loom threaded with metal warp of 32-gauge copper wire and woven in an open weave with a metal weft of the same color and size.

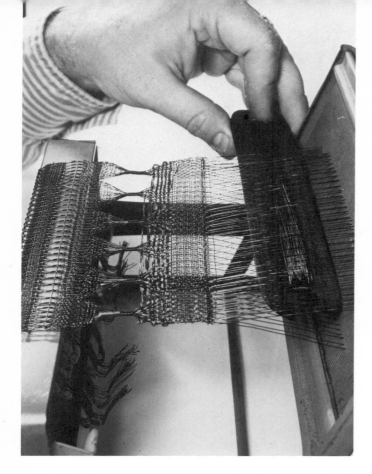

3-25b. The shed is open to receive a flat shuttle wound with the copper-wire weft.

3-25c. After changing the shed, the weft is pushed down firmly with the beater, just as in yarn weaving.

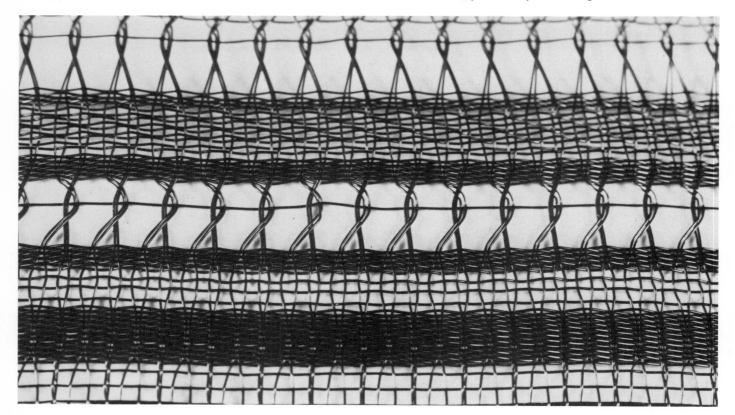

3-26. An evenly spaced warp in a variety of weaves, including plain, basket, leno, and twill.

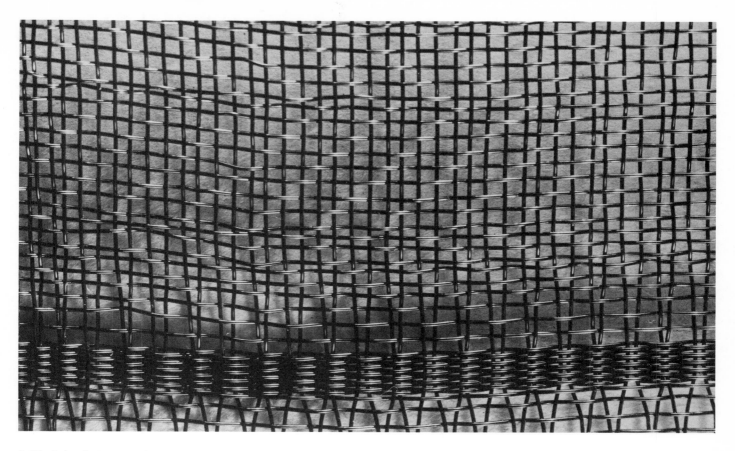

3-27. A herringbone twill in copper wire.

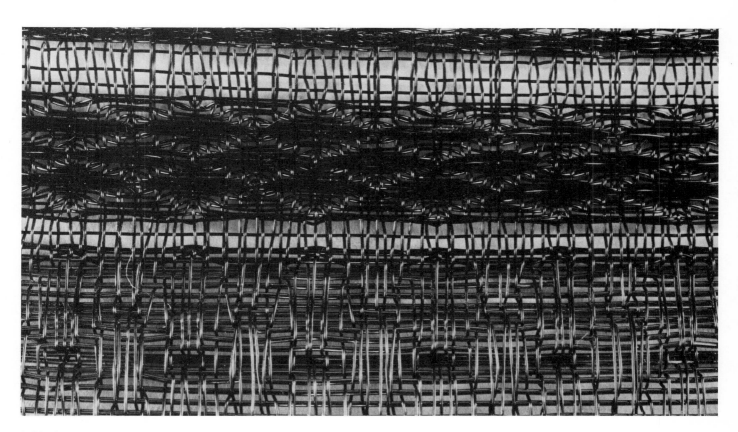

3-28. A series of overshot patterns in copper wire with warp and weft floats. Wire is excellent for such weaves because long floats are not likely to be snagged or deformed in use.

The samples shown here are threaded in various twill patterns, at times with an unevenly spaced warp (3-29 and 3-30). The first inch or two of the warp is woven with heavy yarn or strips of cardboard or metal (as described in the section on cardweaving) to achieve proper warp spacing. In preparing metal wefts, try to avoid bending or crimping the metal unnecessarily. Fine wire can be wound on regular bobbins but heavier wire is best placed on long wooden rug shuttles. Ribbons of metal must also be wound on long flat shuttles to prevent twisting and distortion. Stiffer metal strips may be included in the weft as separate elements (3-30) for unusual surface variety. They are best used precut, and the extending edges can be finished after the piece is off the loom. In weft-faced weaves, as opposed to cardweaving, the weft's size, shape, and color will change the effect greatly, and must be chosen with care. Weft may be wires of various sizes, strips of flat sheet, twisted wires or ribbons of sheet—all in any metal.

Weaving with wire proceeds in the same way as with yarn. The shed is changed and the weft is pushed down with the beater. The front beam of the loom may need padding to keep its edge from bending or damaging the woven metal structure as it is finished and wound up. It is also important to wind sturdy paper or fabric between these layers of finished weaving to prevent any damage to the surface of the metal.

3-30. A basket-weave construction using flat ribbons of copper as the weft.

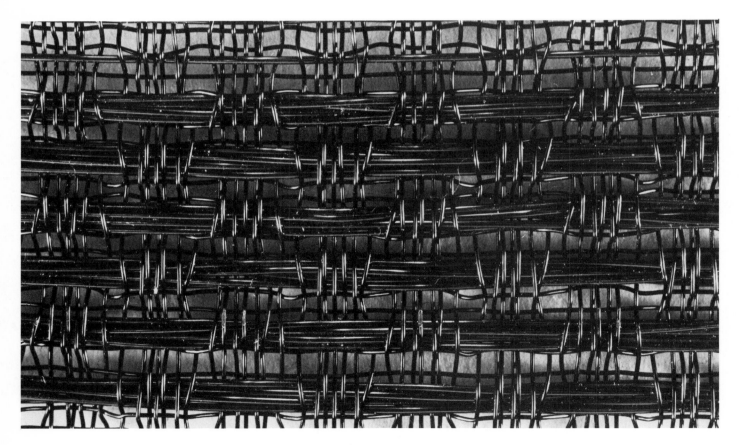

3-29. A single overshot pattern in copper wire with grouped warp and weft floats.

78

With wire, the warp can be merely cut off the loom. Finishing can be by coiling or twisting (3-31), weaving ends back into the structure, or tacking them down with solder. The entire end of the warp may be soldered to a finishing piece of metal if the object calls for it or it may be threaded through holes in a metal sheet or wrapped around a metal rod to provide weight and stability of form.

3-31. Woven on a four-harness loom, this bracelet by Mel Somerowski has a copper weft and a copper warp spaced for dense and open areas. After being cut off the loom, the warp ends were twisted to create loops and buttons as fastening devices. Then the entire bracelet was immersed in a copper plating bath and allowed to plate very slowly so that the wires were made heavier and the piece sturdier while still remaining flexible. The bracelet was finished by plating in gold.

4/Knitting

Knitting is a continuous single-element technique in which a series of loops are interlocked vertically through the repetition of knitting stitches placed on some kind of tool or frame. The new stitches are constantly interlooped into the already existing structure to extend, expand, or compress it. By varying the kinds of loops and their sequence, patterns of many types can be developed. Using tools or frames of a different size can change the appearance of the structure drastically because it changes the size of the loop. The fabric can be produced as a single flat plane, as a relief with areas protruding from the basic structural plane, or as a form totally in the round, such as a tube. Color patterns can be produced by the introduction of additional noncontinuous elements such as wires or yarns.

The instruments used fall into two main categories—needles and pins or frames and rings. The interlooping structure remains basically the same throughout all knitting but the type and scale of tool will change the form produced, either flat or tubular, and also conditions the movements or steps used to produce the basic structure. All of the methods are quite simple once the basic looping movements are understood, but it is a good idea to experiment with each before deciding on a particular approach since similar forms can be produced in a variety of ways.

This chapter describes needle knitting and spool knitting, both very useful and adaptable techniques for working with metal. Most of the processes are similar to knitting with yarn, except in the area of tension. Tension, as well as the size of the tool, determines the size of the loop which is so basic to the structure and appearance of the finished knitting. When knitting in yarn, which stretches, it is usual to determine the amount of tension necessary and then to strive to maintain that tension throughout the piece. Wire has no stretch quality at all, and must therefore be handled in a slightly different manner to keep the loops consistent in size and shape.

A few words about material specifications will help avoid unnecessary difficulties and failures. The material, first of all, must be a very flexible and pliable wire in both size and type of metal to withstand the constant pulling of one loop over another. Very thin gauge wires may be pulled so tightly that they will break apart from the constant manipulation as the knitting progresses unless they are annealed or are very soft in composition. Sterling-silver wire, for example, does not survive the knitting technique well, especially in a small scale, unless it is constantly annealed as it is used, and even then it has a tendency to be too brittle for much wear afterwards. Fine-silver wire, on the other hand, knits very well in almost any gauge, although the finished piece may require annealing to assure continued flexibility. In the same way, copper insulation and electrical wires in a wide range of gauges may be used without any difficulty at all and do not normally require annealing when finished. Fine brass wire must be treated a little more gently than either copper or fine silver since it is somewhat brittle, but it can be used effectively in the thinnest gauges. Eighteen-karat gold wire also knits well with a gentle touch. Some ferrous metals found in small-gauge wire form (iron binding wire, nautical stainless steel) can be used readily and may be plated with precious metals later if another color is desired.

4-1. A large ruffled collar in two colors of 28-gauge copper wire, knitted on #2 straight needles. The basic pattern is ribbing, which is increased by one stitch in each rib section in every other row to produce the ruffled shape. The form is as flexible as fabric.

4-3. A three-dimensional bell motif in 30-gauge fine-silver wire on #1 straight needles. This small form (2 x 4 inches) will be combined with sheet metal, pearls, and gem stones to create a brooch or pendant. The "bell frilling" knitting pattern may be found in *Creative Knitting,* Mary Walker Phillips, page 76.

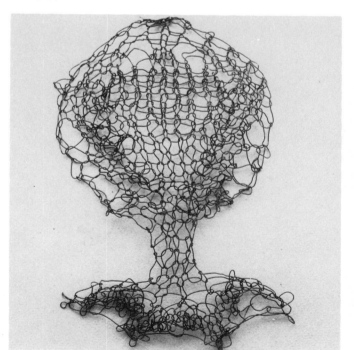

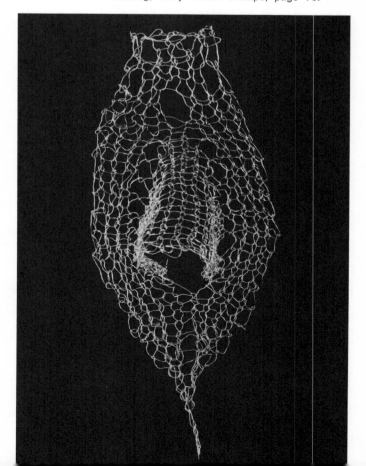

4-2. A tree form in 28-gauge copper wire knitted on #1 straight needles. The basic pattern is the stockinette stitch, which is increased or decreased at both ends of each row to produce the nonrectangular silhouette.

Wire of any gauge can be used for knitting provided it is compatible in size with the instrument used. A wire of 32 gauge can be worked easily on knitting needles size #0 or #1, while a 24-gauge wire will require a much larger needle such as a #8 or #9. (See comparative chart of U.S. and British needle sizes in the Appendix.) Small wires can be worked effectively on needles of many sizes to achieve differences in scale, but the heavier the wire, the larger the initial tool must be, just as with yarn. The same relationship exists with knitting frames and rings where the wire must be compatible both with the total size of the instrument and with the diameter of the individual pins around which the loops are formed. A few test samples of different gauge wires on different sized needles and rings will demonstrate clearly the relationships which are most comfortable as well as most successful. Suggestions can be taken from the illustrations in this chapter.

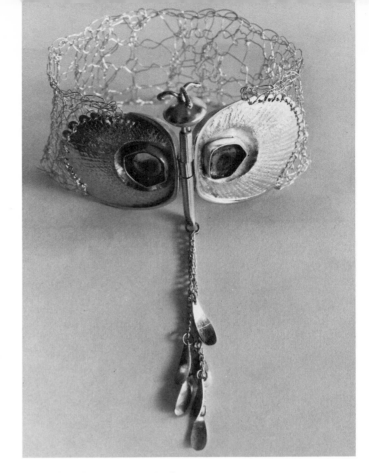

4-4a. *Owl Eyes* (open view), bracelet in 30-gauge fine-silver wire, sewn to a chased silver clasp with a separate piece of wire. The band is knitted horizontally in a lace pattern.

4-4b. *Owl Eyes* (closed view).

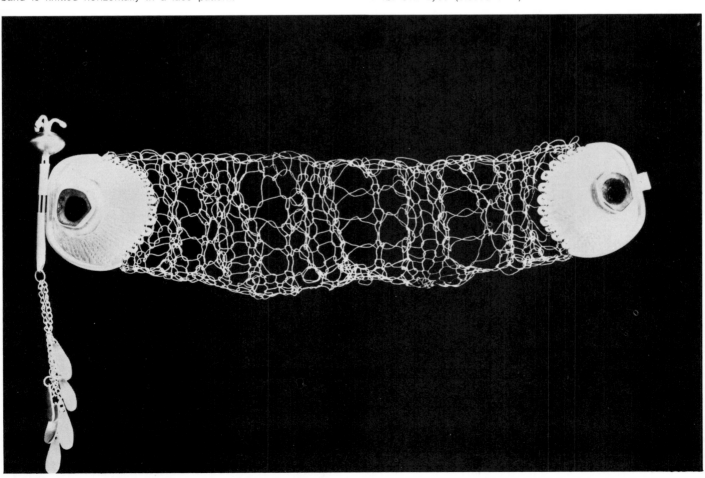

NEEDLE KNITTING

This is the most common form of knitting and offers the greatest versatility in pattern, form, and scale. It can be done on two needles to produce a flat plane; it can be done on four needles to produce tubular forms; it can be done with multiple auxiliary needles to produce various relief and dimensional forms. Instruction in knitting with more than two needles is beyond the scope of this book but can be found in the books listed in the Bibliography.

To begin, a certain number of loops, or stitches, must be cast on a single needle. In wire, this is best accomplished by the simple method of twisting the wire around the index finger to form a loop which is then slipped onto the needle (4-5 a,b,c). It is very important at this stage to hold the wire loosely and to allow some slack between each loop, because without this precaution the first row of actual knitting is almost impossible. Note that the method of casting on illustrated here does not create the knotted effect usual in yarn knitting. With metal, the stitches hold securely enough without the knotting, and the wire is not subjected to that extra stress.

After casting on is completed (4-5d), the second needle is inserted into the first loop (4-5e) from the front left of the stitch to the back right. Then the wire is drawn around the second needle (4-5f) and that loop is pulled through the original loop (4-5g), which is then slipped off (4-5h). This sequence of movements produces the basic *knit* stitch, which may then be repeated in succession across the entire width of the structure.

When the second row is to be started, switch the needles to the opposite hands, so that the second needle is again empty, and begin another row. Once the structure is started, dramatic changes in silhouette can be accomplished by increases or decreases in the number of stitches, or both, making it quite easy to knit both flat and dimensional forms which are neither rectangular nor angular in appearance (see 4-1, 4-2, and 4-3). Of equal, if not greater, importance is the vast range of patterns and textural surfaces which can be accomplished by the simple manipulations of a few basic stitches.

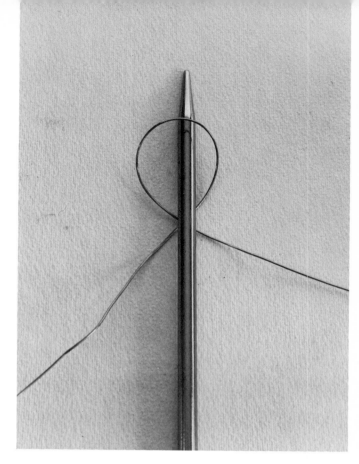

4-5a. Casting on, step one. Make a simple loop of wire and slip it over the end of the needle.

4-5b. Casting on, step two. Wrap the left-hand end of the wire around the needle and pull the resulting loop through the first one.

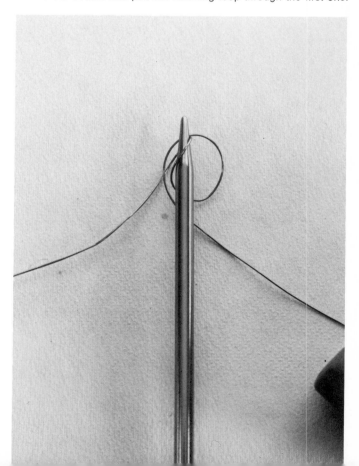

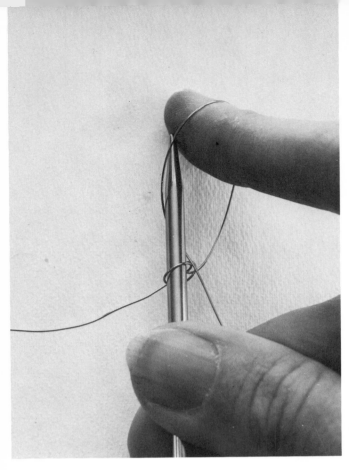

4-5c. Casting on, step three. With the right-hand end of the wire (that which is attached to the spool) twist a loop over the index finger and slip it onto the needle.

4-5g. Knit stitch, step three. Draw the wrapped loop through the stitch by bringing the second needle under and in front of the first needle.

4-5d. Casting on, completed. Cast on as many stitches as necessary, making sure that they are loose enough so that there is an open space between each loop.

4-5h. Knit stitch, step four. Slip the first loop off the first needle, which automatically transfers the new stitch to the second needle.

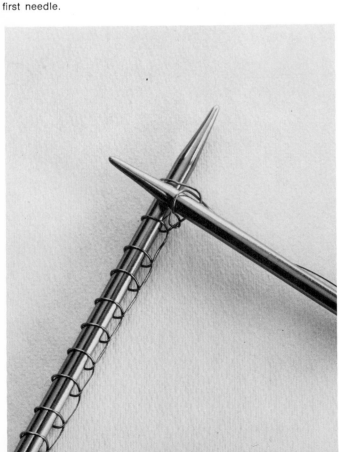

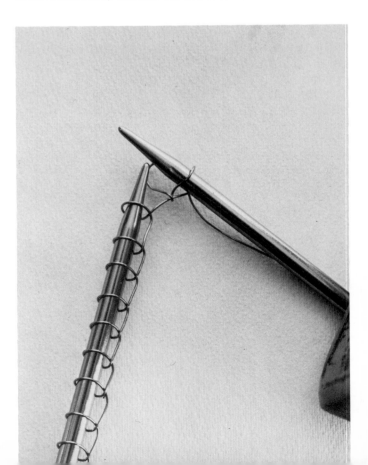

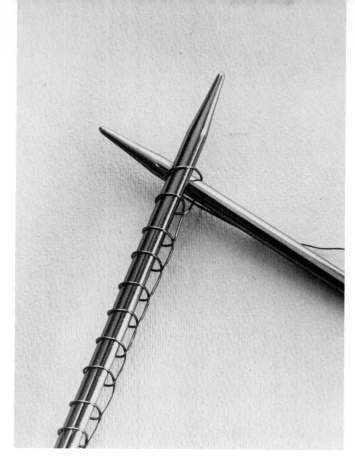

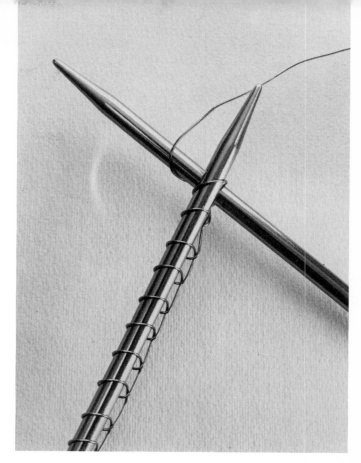

4-5e. Knit stitch, step one. To begin the first row of knitting, insert the second needle into the last cast-on stitch, from front to back, so that it goes behind the first needle.

4-5f. Knit stitch, step two. Wrap the second needle with wire from the spool, so that it goes over the needle from left to right.

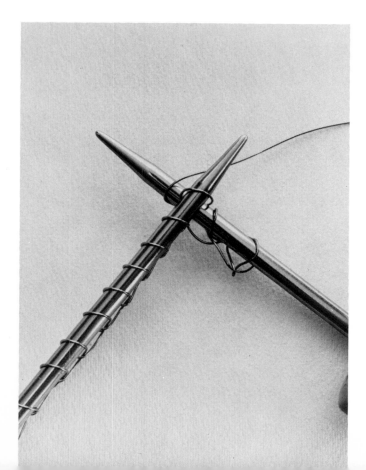

4-5i. Continuing the first row. Draw loops through each successive stitch while keeping the wire slightly loose and the stitches stretched out along the length of the needle. When the row is completed, transfer the second needle to the other hand, and begin again from step one.

When knit stitches are used in every row, the pattern is called plain knitting, or garter stitch (see 4-12). The other basic stitch, the *purl* stitch, is made in exactly the same way as the knit stitch first described, except that the needle is inserted from the back right of the stitch to the front left before the yarn is wrapped around it. A combination of knit and purl stitches produces many variations in pattern, from the well-known stockinette stitch to open lace knits. A list of abbreviations for following printed knitting patterns may be found in the Appendix.

The basic stockinette stitch is comprised of regularly alternating rows of knit and purl, and can be combined with other patterns to produce a wide variety of effects. In the gold cuff (4-7), stockinette stitch with subtle changes in scale has been accomplished in metal by periodic changes in needle size (from #8 at the bottom to #2 at the top) although in yarn the structure could have been made looser or tighter by adjusting the tension as well as the needle size. The overall form is achieved by gradually decreasing the number of stitches on either end of each of the separately knitted halves. There are several ways of increasing and decreasing, and details can be found in knitting books. One can increase the number of stitches in a row by knitting twice into a single stitch before dropping it off the needle or by casting on a new stitch between two stitches. Decreasing is done by inserting the second needle through two adjacent loops on the first needle, and knitting only one loop through them both.

4-7. A cuff knitted in the stockinette stitch in 30-gauge 18-karat gold wire. It was done in two separate pieces on straight needles of various sizes: #8 at the bottom with double wire, #6, #5, and finally #2 for the straight upper section. The size of the needle is extremely important in knitting with wire, because the tension, and thus the size of the loops, cannot be easily adjusted in any other way.

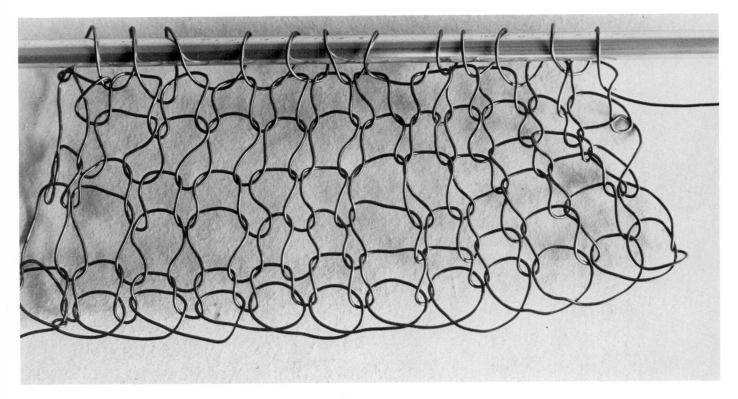

4-6. A sample of knitting in 24-gauge copper wire, done on #8 needles in the stockinette stitch, which alternates one row of knit stitches with one row of purl stitches.

Stockinette stitch is combined with ribbing in 4-8. Ribbing is the basic pattern that results from a regular alternation of knit and purl stitches. The most common form of ribbing is *knit two, purl two,* although many other variations may be used.

Knitted lace patterns are especially beautiful when done in wire. The delicate tracery of the openwork is more discernible than it would be in yarn, and the light-reflective quality of the wire lends brilliance to the entire structure. All of these patterns are based on alternations of increases and decreases, combined with stitches which are passed from one needle to the other without being knitted to create a series of holes and elongated loops which cross and interlock in endless variations. To create a "hole" for a lace pattern, a series of these operations is performed together. For example, the wire is wrapped around the needle without the needle being inserted in a stitch (this *yarn over* effectively increases the number of stitches); a stitch is passed from one needle to the next without knitting (this *slip stitch* creates the hole); a regular stitch is knitted; the slipped stitch is drawn back over the knitted stitch (this *pass slip stitch over* locks the hole in place); finally two stitches are knitted together to keep the number of stitches in the row even (this decrease makes up for the previous increase supplied by the *yarn over*).

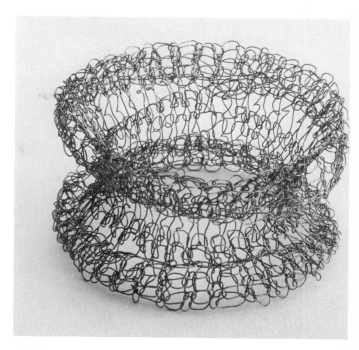

4-9. A tubular bracelet form of 24-gauge copper wire, knitted in the round on four double-pointed #8 needles. The top and bottom are in stockinette stitch and the center uses several rows of double throws, that is, wrapping the wire twice around the needle to produce a longer stitch and a more open structure.

4-10. A section of a rectangular bracelet panel knitted in 30-gauge fine-silver wire on #2 needles. The heavier border areas use a double wire in stockinette stitch and the central section is a single wire in a lace faggot pattern, which may be found in Phillips' *Creative Knitting,* page 68.

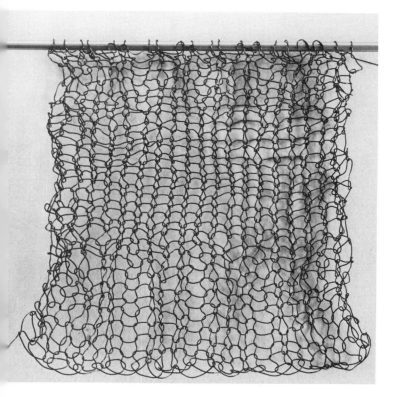

4-8. A sample of knitting in 28-gauge copper wire, done on #2 needles in both the stockinette stitch (center portion) and a ribbed pattern of *knit 2, purl* 2 (top and bottom).

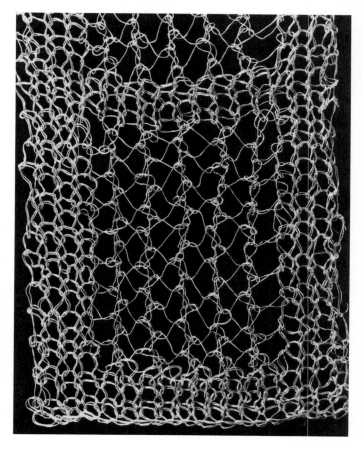

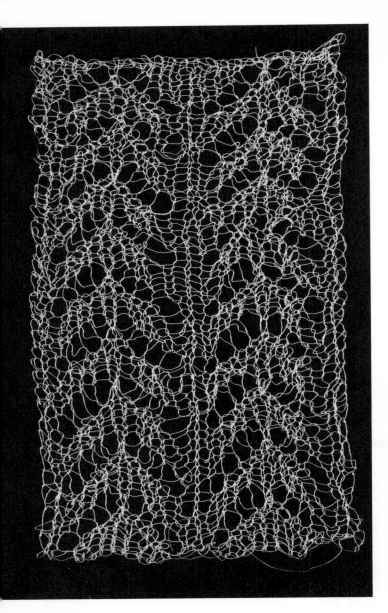

The lace patterns may be used alone (4-4) or incorporated into other, more densely knitted patterns (4-10 and 4-12). In many instances, areas of denser knitting provide a visual and structural frame for the more delicate lace patterns and eliminate the need for support from nonknitted structures. The same principles used in forming lace patterns are used to produce three-dimensional motifs (see 4-3) that offer exciting possibilities for the development of surfaces in both small-scale jewelry and large-scale wall hangings or screens.

In yarn knitting, the last row of stitches is secured in a way similar to the first row put on the needle. The process is called casting off. With metal, casting off must be done as loosely as casting on in order to maintain the proper width of the piece. Other methods of ending may be used in place of casting off because the wire loops can easily be secured by twisting on themselves or by securing to some finishing piece. There are many possibilities, such as running a wire through all the loops or chaining them off as in crochet.

4-11. A rectangular panel, 6 x 10 inches, knitted in 30-gauge fine-silver wire on #1 needles. The basic lace pattern, which may be found in *American Needlework Book of Patterns* #43, Pattern #6, page 28, is used twice in a symmetrical repeat.

4-12. A rectangular bracelet panel knitted in 30-gauge copper wire and 30-gauge fine silver wire on #2 needles. The border is done in a garter stitch (plain knitting) with double wire—one of each metal—and the center is done in silver in a lace pattern that may be found in the *American Needlework Book of Patterns #43,* Pattern #10, page 28.

SPOOL KNITTING

This form of knitting, common among children, is done on a spool with four nails or pins placed around a central hole. Sometimes it is called bobbin knitting, ring knitting, or French knitting. I can remember making literally miles of what were called "horse reins" by my contemporaries, although I cannot recall even a hint of what these masses of tubes might have been used for except dollhouse rugs, potholders, and small hats. Such a device is still marketed as a "children's knitting set," "knitting hobby," "knitting Nancy," or "horse-rein spool knitter."

4-14. Spools for knitting. *From left to right:* a wooden cylinder with six wooden pegs, a wooden spool with six nails, and a molded plastic tube with four pins. This sort of tool is usually found in the toy department, not the craft department.

4-13. *Golden Reins,* spool-knitted tube, 30 inches long, in 18-karat gold capped at the ends with metal tubes and antique carnelian beads. The necklace is simply tied around the neck in a loose knot.

4-15. Samples of small knitted tubes from Yemen. The tube on the left is a single-loop structure using six pins; the one on the right is double (see 4-20) and done on four pins to produce a square cross section.

89

4-16. Four samples of knitted tubes in fine-silver wire. *From left to right:* 26-gauge wire knitted on eight pins of a three-inch diameter plastic ring with wire drawn across rather than around each pin; 30-gauge wire knitted around five pegs on a wooden spool; 30-gauge wire drawn across six pins and then twisted by hand; 30-gauge wire knitted around five pegs and stretched by pulling through a drawplate (same as second from left).

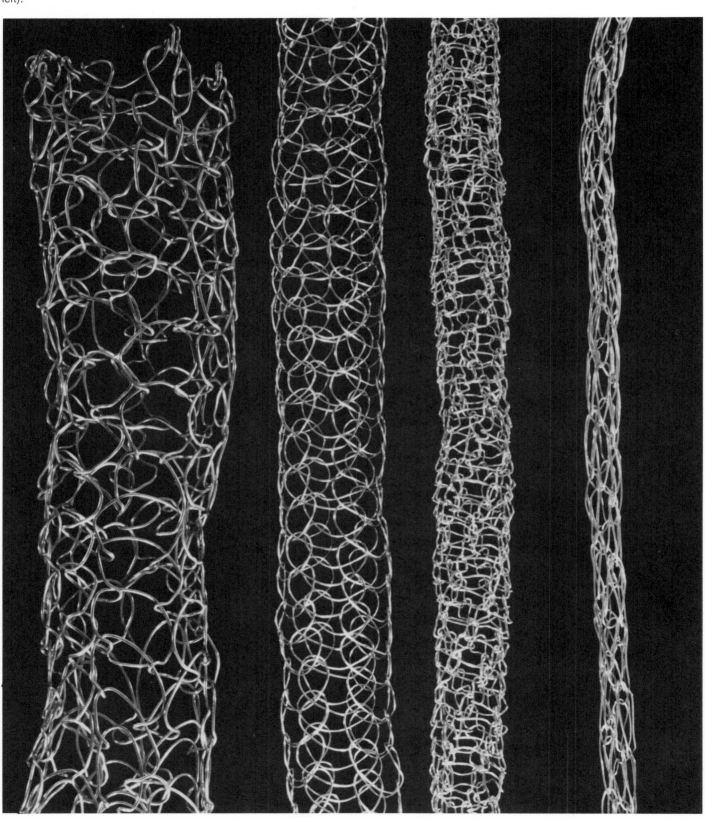

There are many variations of the knitting spool now available on the market, some in the form of rings (see 4-17), but they are almost always to be found in the toy department rather than in the yarn supply shop. All of the commercial ones are made of plastic, which is actually preferable for working in wire because it causes less damage to the loops as they are pulled over the pins. A very small spool can be constructed at home from an empty spool of thread with nails placed around the opening. Rings can be made in any dimension by cutting holes in plywood sheets. Where possible, it is good to substitute small wooden pegs for the usual nails to reduce wear on the wire and to control more accurately the size and shape of the loops. The tool for lifting the loops over the pins should also be as nondamaging to the metal as possible—a plastic needle or crochet hook or a smoothly pointed and polished straight metal rod mounted in a wooden handle.

My first attempts at using this method of knitting with wire were done on the same simple four-pin spool I remembered from childhood, but the results were very limited in both scale and pattern. Many variations can be made by increasing the size of the spool, which increases the diameter of the tube, and by increasing the number of pins, which makes for a denser structure (see 4-19). Changing the size and shape of the pins themselves produces loops of different dimensions. The method of knitting can be varied too. The rows of loops can be doubled (see 4-15), pins can be skipped over, or loops can be moved sideways to give a spiral.

4-18a. Casting on, step one. Tie a length of yarn in a simple slip knot at one end. The yarn is used, before the metal, to avoid distorting the metal structure as it is pulled through the ring.

4-18b. Casting on, step two. Place the knot over one pin, with the short end through the center of the ring.

4-17. Two variations of plastic knitting rings sold in children's knitting kits. The one on the left has 16 pins; the one on the right has 18. The polished metal rod set in a wooden handle is the tool used for lifting the loops over the pins.

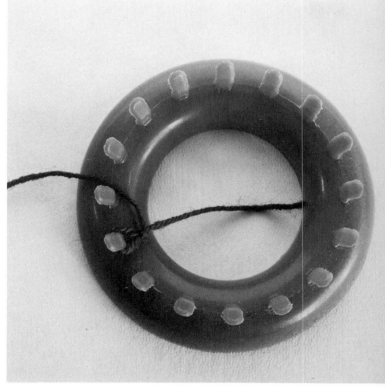

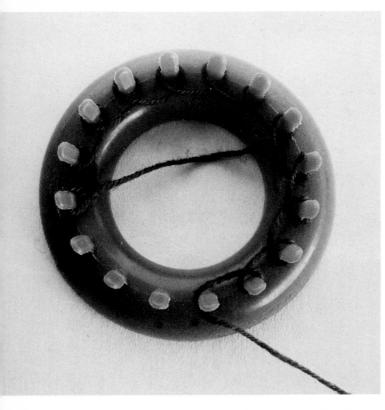

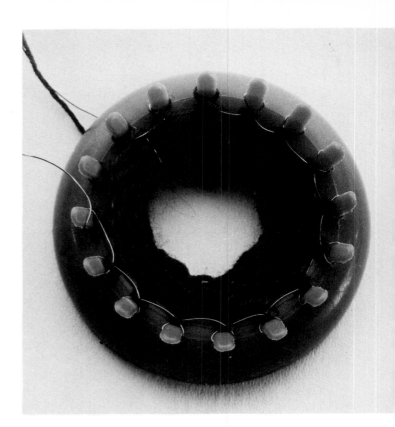

4-18c. Casting on, step three. Wrap the yarn around each successive pin in a clockwise direction, with the crosses of the loops facing the inside of the ring.

4-18e. Knitting, step two. After an inch or two of fabric is knitted, the wire is added by looping it together with the yarn for several stitches, then dropping the yarn and continuing with the wire alone.

4-18d. Knitting, step one. Loop the yarn around a pin, and lift the cast-on loop over the new loop. Continue looping around successive pins, lifting the first row up and over the second row. As knitting proceeds, constantly pull the tube downward into the center of the ring.

4-18f. Knitting, step three. Constantly pull the structure down into the center of the ring as the work progresses. When the required length is reached, finish off as with regular knitting or simply run a wire through the loops to hold them.

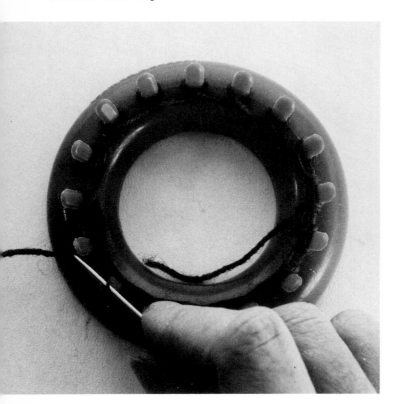

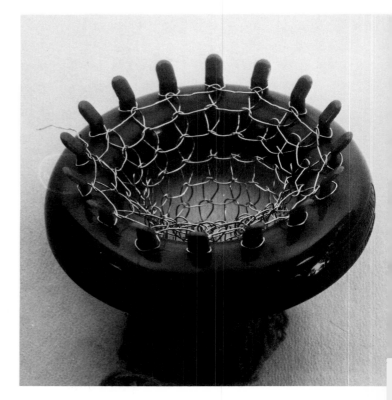

4-19. A necklace knitted in 26-gauge fine-silver wire on a 16-pin plastic ring. The finials and clasp are constructed in sterling-silver sheet and inset with antique glass slabs in iridescent blues and greens.

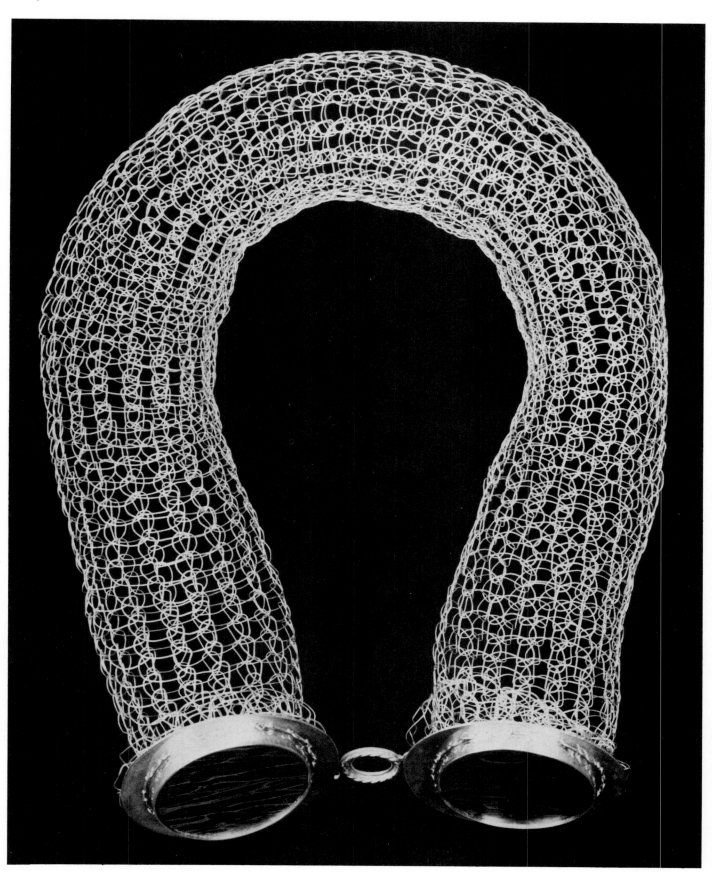

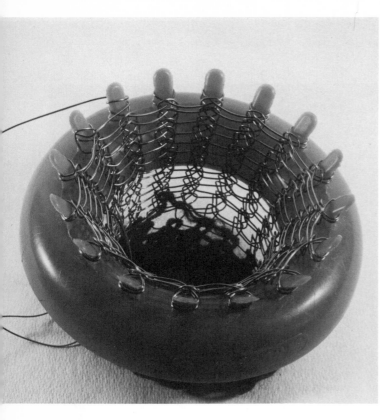

4-20. A variation in which double loops are used to produce a denser structure. Two rows are kept constantly on the pins; when a third row is added, the bottom row is lifted up and over two rows.

4-21. A Yemenite sample of double-loop knitting with the tube flattened. A vertical stripe pattern is very prominent.

4-22. A variation on spool knitting is done without spool or frame, and creates a flat or tubular structure. The crochet hook reaches back into multiple rows of loops, allowing greater density and pattern variation in the knitted structure. The sample in 4-21 may have been made this way.

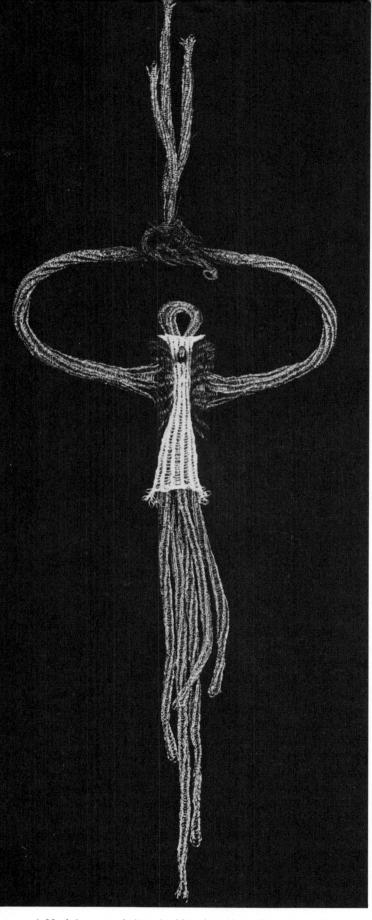

After the tube is completed, a deliberate flattening of it—by hand or rolling through a rolling mill—creates a less crushable structure from the invariably delicate tube. Stuffing tubes with fabric, leather, or additional tubes in different colors of metal would also increase the structural strength and allow for a more extensive use of spool knitted elements as single units or in multiple-element structures. Almost all of the historical and contemporary examples of this technique, such as the samples of Yemenite work shown here and in Chapter 1, are in the form of chains and neckpieces, since these objects are relatively protected from crushing.

Another method of producing knitted structures similar to those of spool knitting uses only a crochet hook, without a frame. This technique is clearly seen in 4-22, where a nontubular form is being made. The technique can be done in tubes or flat, and the forms can be expanded or contracted by increasing or decreasing the number of stitches. This technique allows more variability in pattern changes and density of structure than regular spool knitting. These changes are made by inserting the crochet hook back into the existing structure in the second, third, fourth, or fifth row from the top. The new loop is pulled forward to form the next row, overlapping the rows in between, which greatly increases the density of structure. This technique was probably the one used to produce the Yemenite samples in 4-15 and 4-21, as well as the knitted neckpiece by Mary Lee Hu.

4-23. A large neckpiece by Mary Lee Hu in fine silver and coated copper wire. The entire tubular structure is fashioned without the use of a spool, using the technique shown in 4-22.

5/Crochet

Crochet, like knitting, is a single-element structure in which loops are interlocked in a continuous manner. In crochet, the structure is locked in one direction as it is created in the other, and only one loop remains on the hook at the completion of each stitch. Crochet is a technique which can move freely in any direction, and is well suited to both openwork patterns and three-dimensional forms because it is possible to add to the structure at any point. There are virtually no restrictions in the way stitches can be combined or the directions in which they can be assembled.

Crochet is an excellent technique for adaptation to working in metal because metal offers more support than yarn and the most lacy structure or the most varied shape can be readily adapted to jewelry and ornamental garments. Colors and sizes of wires can be interchanged freely, and areas can be built over or out of any surface, allowing the development of relief and dimensional forms. The wire must be pliable however, since the metal is constantly subjected to a twisting and tugging motion. As with knitting, no special tool other than a crochet hook is needed, although experimentation will be necessary to decide how large a hook is appropriate for the gauge of the wire. Tension must be carefully controlled, and if larger loops are desired, a larger hook must be used. A comparative list of crochet hooks is included in the Appendix, since sizing differs in the United States, Great Britain, and Europe. No specific recommendations can be given here about what size hook to use with each gauge of wire, since the choice can be so varied. For the very thinnest gauges one of the steel hooks that are used for fine thread may be appropriate and for the thicker gauges, the aluminum and perhaps the jumbo plastic hooks should be tried. It is possible to work only with the fingers when using large scale or particularly stiff material in simple stitches.

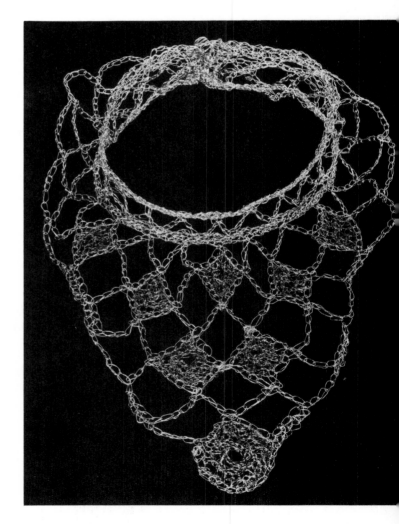

5-1. Crocheted necklace by Ana Castro in 26- and 28-gauge copper wire. Simple crochet stitches such as netting of chain stitch create the structure. The fastening is a metal button and simple crocheted button loop at the back of the neck.

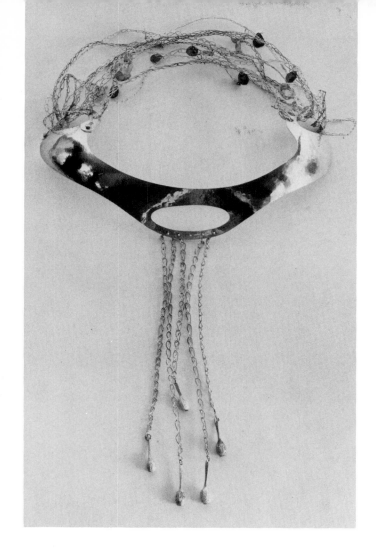

All crochet is started from a foundation row of simple *chain stitch* (5-6a,b), which can be expanded to form intricate linear patterns (see the necklaces in 5-1 and 5-2). The more complex stitches are built up from the simple looping movements of the chain stitch, and can be used to form denser structures and more elaborate patterns. The *single crochet* (5-3) is the basic unit from which the *double crochet* (5-4) and higher stitches are formed. The hook is wrapped an increasing number of times for the triple and quadruple stitches, and with each wrap the stitch becomes higher, forming the equivalent of several rows in one passage, in an increasingly open structure (see 5-7).

5-2a. Necklace in silver and gold (back view). The fastening is two rivets that fit into slots at the left of the molded silver sheet.

5-2b. Detail of front view of necklace. The chains of single crochet are in varying gauges of sterling, fine silver, 14-karat gold, and 18-karat gold wire, in gauges from 30 to 22, and were crocheted with hooks of varying sizes.

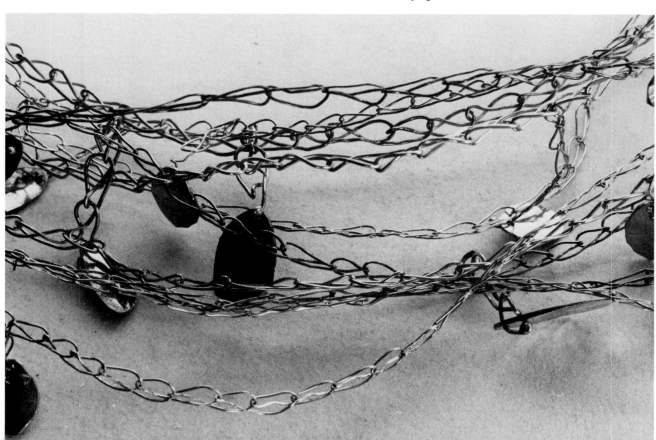

A brief explanation of chain stitch, single, and double crochet should suffice to begin some practice samples. More complete explanations and diagrams can be found in almost any instruction booklet in yarn shops and needlework departments, as well as in numerous books currently available on the subject. A list of crochet abbreviations, which will aid in adapting some of the numerous printed patterns found in books and magazines to metal, may be found in the Appendix.

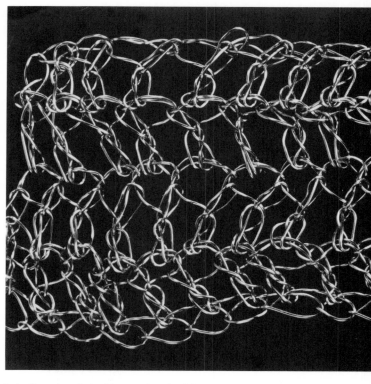

5-4. Sample of double crochet in 26-gauge coated copper wire, done on a #9 hook. Note that each row of the double crochet is twice as high as the single crochet, and the stitch has a more open effect.

5-3. Sample of single crochet in 24-gauge coated copper wire, done on a #E hook.

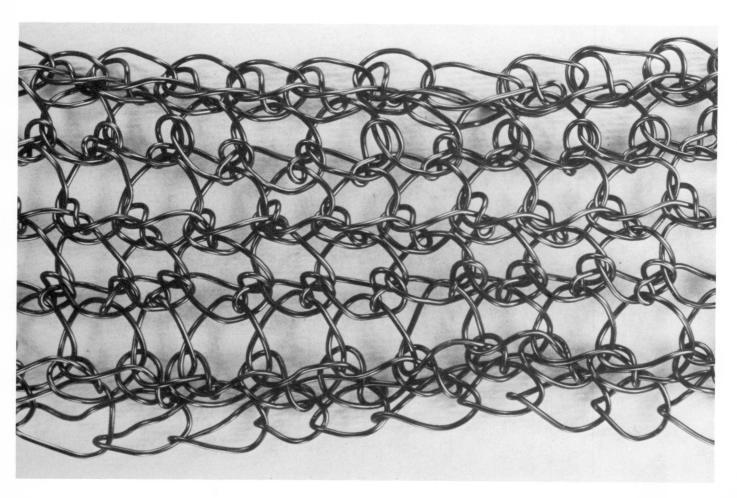

5-5. Crochet equipment consists of crochet hooks and frames (for hairpin lace) of varying sizes. *From left to right:* aluminum crochet hook, size F; aluminum frame, 2 inches; three steel crochet hooks, sizes #9, #5, and #1; an adjustable frame with plastic supporting ends; and a steel frame, ¾-inch wide. Sizing systems for crochet hooks vary in other countries, and a comparative chart may be found in the Appendix.

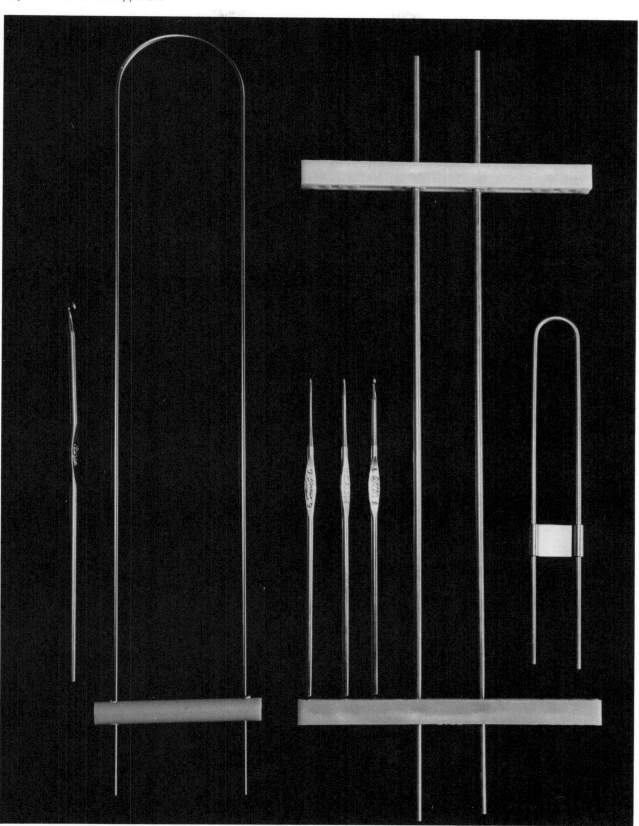

The slip knot that begins the foundation chain may be difficult to form with metal. Instead, twist a loop of wire around the hook, draw an additional loop through it, then pull firmly on both ends of the wire. To make a chain stitch in wire, hold the short end of the wire at the base of the knot with the thumb and index finger of the left hand and hook the long end of the wire from underneath. Draw this through the existing loop, so that a new loop is on the hook. Note that in crocheting with yarn the yarn is hooked in a slightly different way, which is described as "wrapping the hook" or "yarn over." This creates a little extra twist in the material which is not necessary in metal. However, the final result is very similar. Repetition of this movement forms the chain of stitches.

The single crochet is made by working back into the loops of the chain (5-6c through f) and the double crochet and higher stitches are accomplished in the same way after first wrapping the hook with yarn (5-8). When beginning the first row after the foundation chain, the hook is inserted into the second or third chain stitch from the hook for single crochet and the third or fourth from the hook for double crochet to allow for the height of the stitch. Each time the work is turned to start a new row, extra chain stitches must be made for the same reason.

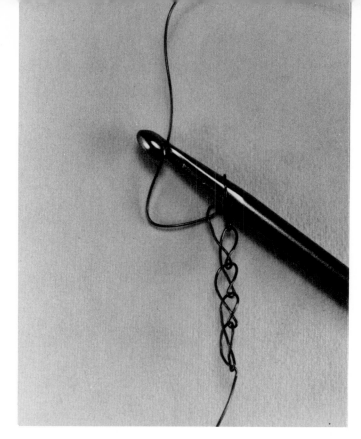

5-6a. Chain stitch, step one. After making an initial slip knot on the hook, catch a loop of wire as shown. Note that it is easier to catch the wire this way, rather than wrapping it around the back of the hook as is traditionally done with yarn. The sample is 24-gauge copper wire done with aluminum hook size E.

5-6d. Single crochet, step two. Draw the wire through the loop, and now you have two loops on the hook. Catch the wire a second time.

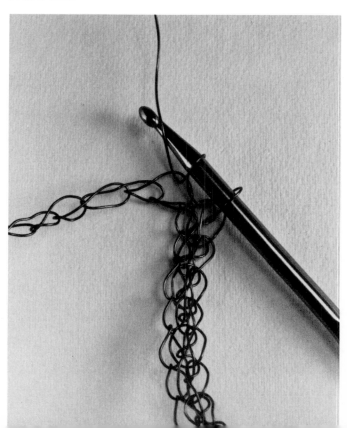

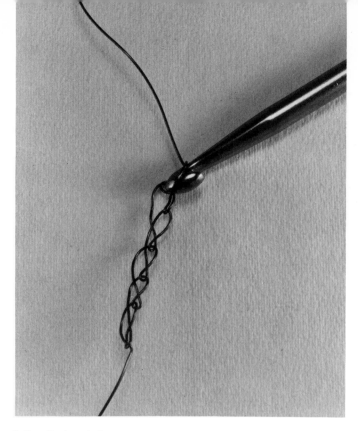

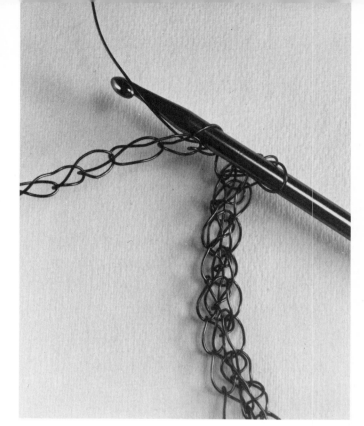

5-6b. Chain stitch, step two. Draw the wire through the loop on the hook. Continue with these two steps to create as long a chain as necessary. The loops should be kept large and loose when working in wire. Five complete stitches are shown here, as well as the one in progress.

5-6c. Single crochet, step one. When the chain is as long as desired, begin to work back into the loops of the chain as shown. Insert the hook in a loop, and catch the wire in the same way as before, by passing it to the left and under the hook. Note that the first insertion will have to be in the second loop from the one on the hook to allow for the height of the single crochet.

5-6e. Single crochet, step three. Draw the wire through the first loop on the hook.

5-6f. Single crochet, step four. Draw the same wire through the second loop on the hook. In practice, step two and three are accomplished in one movement to complete the single crochet.

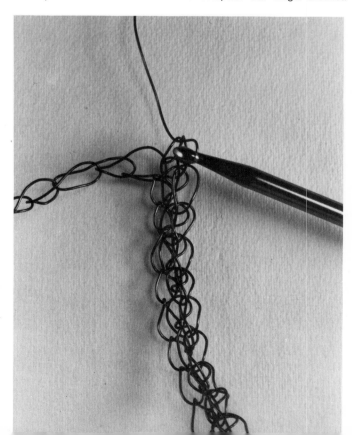

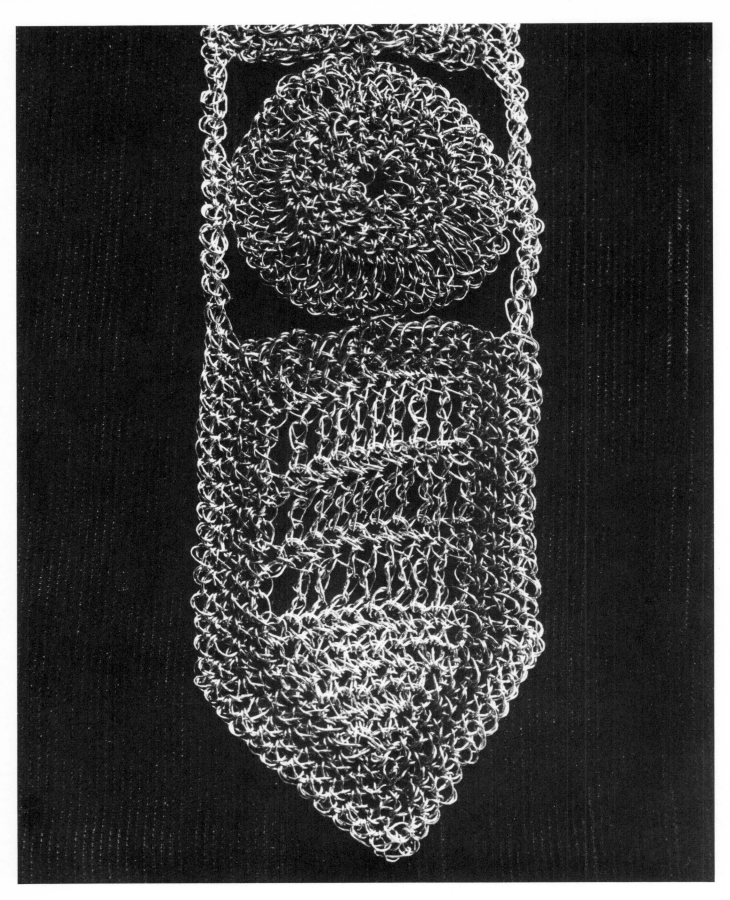

5-7. Detail of crocheted belt by Christine Brown. Various colors of coated copper wire are used in a series of double and triple crochet. This produces a dense and self-supporting structure.

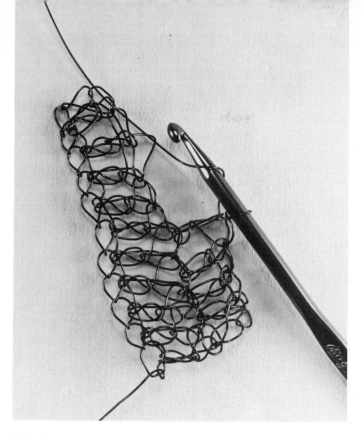

5-8a. Double crochet, step one. Wrap the hook as shown.

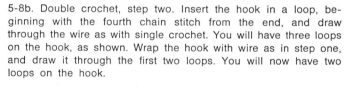

5-8b. Double crochet, step two. Insert the hook in a loop, beginning with the fourth chain stitch from the end, and draw through the wire as with single crochet. You will have three loops on the hook, as shown. Wrap the hook with wire as in step one, and draw it through the first two loops. You will now have two loops on the hook.

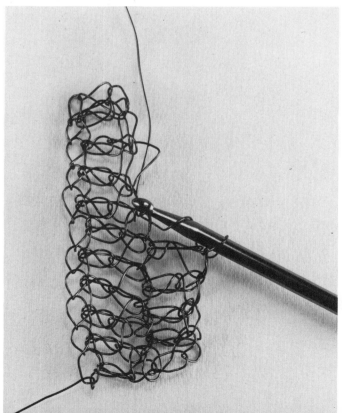

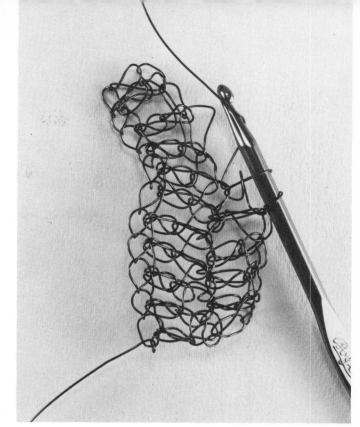

5-8c. Double crochet, step three. With two loops on the hook, wrap the hook with wire again.

5-8d. Double crochet, step four. Draw the wire through the remaining two loops to complete the double crochet. A triple crochet is made in the same way as a double crochet except that the hook is wrapped twice to begin with before the loops are worked off two at a time.

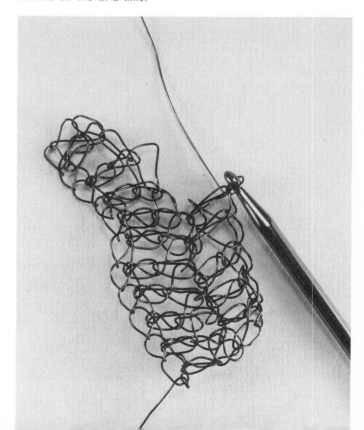

HAIRPIN LACE

Hairpin lace, or hairpin crochet as it is sometimes called, is worked with a crochet hook on a strong U-shaped frame with a brace across it, to produce a band of loops which are the width of the diameter of the frame. The technique is traditionally used to make laces and lace edgings of very delicate scale, and perhaps the first frames really were hairpins. In recent years, larger frames have been used widely with large-scale yarns to make afghans, stoles, and other loose garments.

Hairpin lace is ideally suited for use in pliable wire of almost any dimension because the heavier body of the wire permits the intricate patterns of the technique to show clearly. With wire, the loops can be bigger, not sagging as much as yarn does, and this means that broader bands can be made if wider frames are improvised with a coat hanger or something similar.

The frames are available in a wide range of sizes, types, and materials (see 5-5). The U-shaped frames are not adjustable in size, but are sold in steel and aluminum, ranging from one-half inch to three inches.

The large frame on the right is adjustable by moving the vertical rods into various holes in the plastic elements at top and bottom, producing widths from one inch to four inches. When working on narrow strips, it is simpler and more comfortable to use a fixed frame of the desired size, since the large adjustable type is rather clumsy to handle.

As in spool knitting, the technique will produce an infinite length of continuous structure. The finished portion is slipped off the end of the frame as necessary. There are a few variations possible in making the structure. They alter not only the appearance but the way the strip of lace behaves after its removal from the frame. First of all, the size of wire and hooks may be varied considerably within each frame to produce different effects. Single, double, and even triple crochet may be used as the central stitch, and the insertion point of the hook produces either a dense or a more open knotted appearance (5-11), and changes the vertical expandability. The placement of the central stitch can also be manipulated to undulate from side to side within the width of the strip (5-12).

5-9. A necklace in hairpin lace, made of 30-gauge fine-silver wire and 32-gauge dark copper wire. Seven strips of varying widths and lengths are joined through interlocking to produce a 4-inch wide collar which encircles the neck.

Except for supplementary crochet used to join or reinforce strips (5-11b) after they are off the frame, the only crocheting done in hairpin lace is the central stitch in each loop around the frame. The technique is simple —the same movements are made each time. The wire is first tied around the frame in a single knot and then tied again in the center of the frame to form two loops (5-10b). To proceed, a spool of wire is brought around the right rod of the frame, and then a single crochet is formed (5-10c through g). In order to turn the frame, which is the next step, the hook must be flipped through the frame or removed and replaced in the same loop on the other side of the frame if the frame is too small for the hook to go through. The frame is given a half turn to the left, by rotating the right side toward the crocheter, which adds a loop to the right rod. Another single crochet stitch is made, the frame is turned again, and so on until a strip of the required length has been made. At the end of the strip, the wire is cut three or four inches long and pulled completely through the last loop remaining on the hook, as in regular crochet.

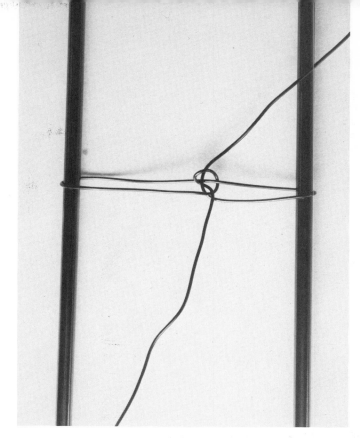

5-10b. Hairpin lace, step two. Knot the short end (at bottom) around the entire loop, as shown, to divide it into right and left loops.

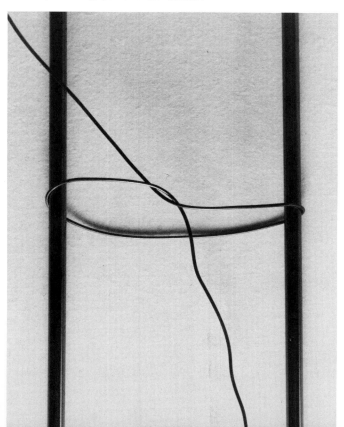

5-10a. Hairpin lace, step one. Tie the wire around the frame, with the crossing point in the middle.

5-10c. Hairpin lace, step three. Wrap the spool end of the wire around the right side of the frame, insert the hook into the left loop, catch the wire as shown, and draw it through the loop.

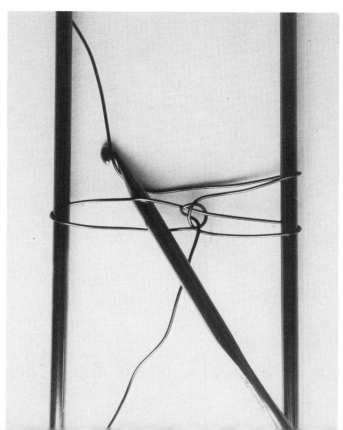

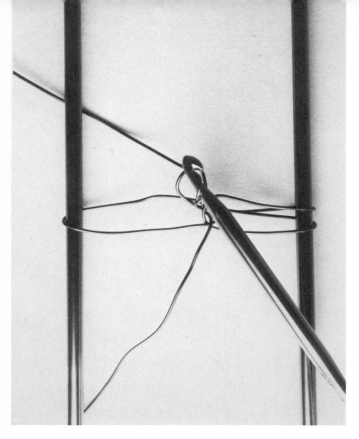

5-10d. Hairpin lace, step four. Draw the wire through the new loop on the hook to secure it with a chain stitch.

5-10e. Hairpin lace, step five. Make one half turn of the frame so that the right side rotates toward you to become the left side. Flip the hook through the frame so that it is again in position, or, if working with a small frame, remove the hook while turning the frame and then reinsert it. Here the hook is inserted under the entire left loop, but it also may be inserted into the loop (see 5-11a).

5-10h. Hairpin lace, to continue. Repeat steps four, five, and six until a band of the required length is formed. Notice the pattern created by the insertion of the hook under the entire loop.

5-10i. Hairpin lace, removed from the frame. This band is one-inch wide, and is made of 30-gauge copper. A steel hook, size #9, was used.

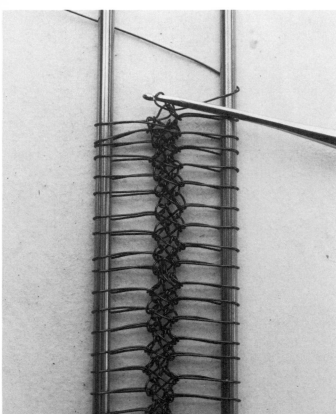

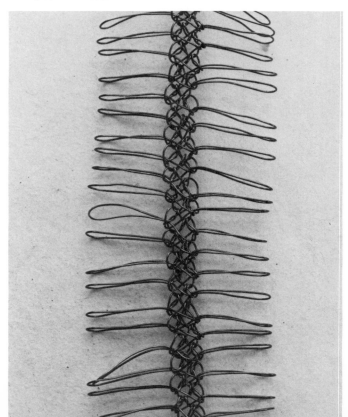

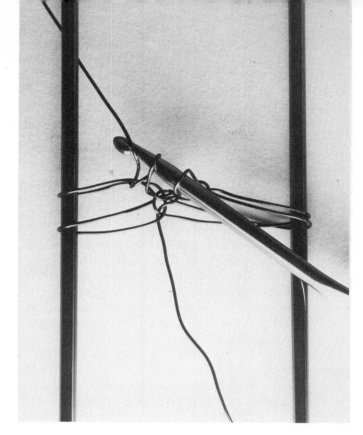 

5-10f. Hairpin lace, step six. Catch the wire again, as in step three.

5-10g. Hairpin lace, step seven. Draw the wire through the two loops on the hook to complete a single crochet. You are now ready to turn the frame again, as in step five. The frame must always be turned in the same direction, and the hook must always be inserted in the left-hand loop.

5-10j. Hairpin lace, to join bands. A simple interlocking method is used here. Each loop is drawn under the next loop on the opposite side, in succession, until the top loop is reached; this one can be secured by twisting or tying.

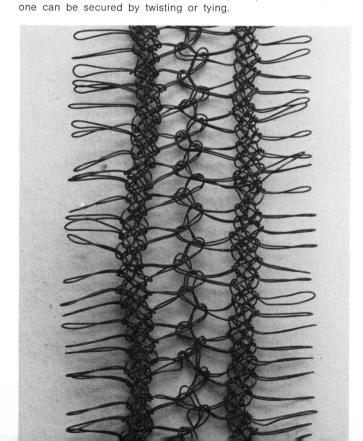

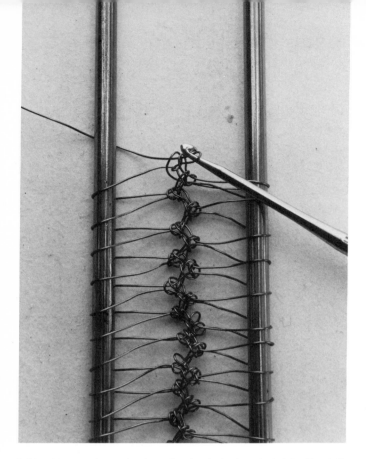

The strips are not sufficiently sturdy as they come from the frame to be used as single-element constructions. They must be given additional support either by attachment to a frame of wire or sheet metal, by enrichment with further crochet chains along the edges, or by joining several strips together. Even in yarn, the strips are joined together in series to produce the finished fabric. There are several methods of joining one strip to another, and they all have their own surface qualities and flexibility. The most common practice is to interlock groups of two or three loops from the inside edges of adjoining strips (see 5-10j). This may be accomplished by pulling them through with a crochet hook. Another method of joining is to crochet either a chain stitch (5-11b) or more complex crochet stitch with another wire—it may even be a different color—so that it catches the adjoining loops one at a time. In all instances, however, the additive method is best used where the strips remain in a straight, vertical, or horizontal pattern. When curved forms are desired, the interlocking method is the most satisfactory because it allows for extensive expansion and contraction. The necklace in 5-9 uses the interlocking structure to join seven strips of slightly varying lengths and widths that expand on their lower edge to produce the necessary curved form.

5-11a. Pattern formed when the hook is inserted into the left-hand loop, as traditionally done with yarn.

5-11b. Three bands made in this fashion are joined by chaining: put a slip knot on the hook, then insert the hook into the first loop, make a chain stitch, insert into the loop on the opposite side, make a chain stitch, and so on.

5-12a. The hook has been inserted into the left-hand loop, but the stitch has been moved slowly from side to side in an undulating pattern.

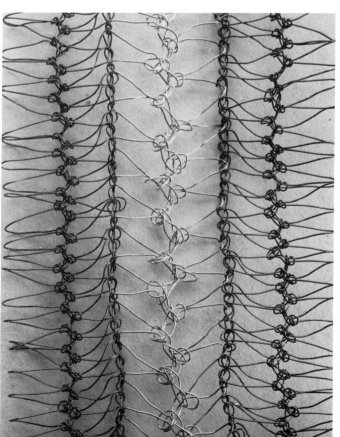

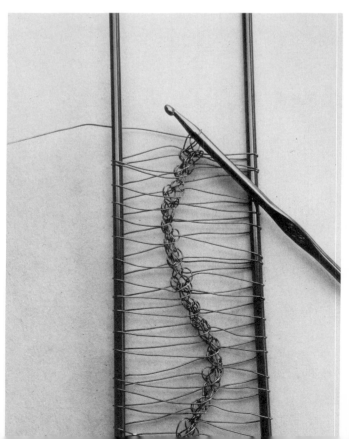

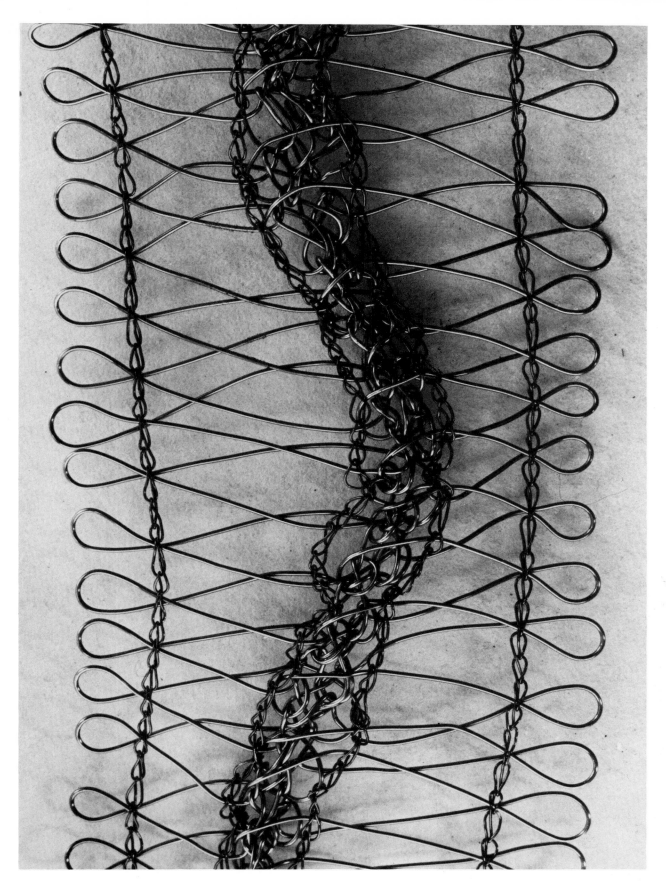

5-12b. As part of the creation of a bracelet in 26-gauge copper wire, structural strength has been added through the use of supplementary chain-stitch elements, two on the outer edges and two around the central core.

6/Braiding

Braiding is one of the more difficult techniques to describe, not because it is so complex—almost everyone knows how to make the common three-strand structure used in braiding hair—but because its terminology has become so confused. Irene Emery, in her book *The Primary Structure of Fabrics,* defines it most clearly as a simple interlacing of a single set of elements without any type of link, loop, wrap, or knot. It is differentiated from weaving, which is also an interlaced structure, in two ways. First, it uses only one set of elements to form the woven structure, that is, the warp serves as weft. Second, the interlacing is in a diagonal or generally oblique pattern, with changes of direction normally occurring only at the edges of the fabric where the elements turn back in the opposite direction.

Braiding is frequently given a number of other names, such as plaiting (the preferred term in Britain), webbing, and interlacing. Many weavers make complex patterns in braiding but call it finger weaving, a system in which the warp is attached at only one end, and a shed is created with the fingers for the weft strand instead of weaving it over and under in sequence. If a separate weft element is used in finger weaving, it is not technically braiding, and frequently finger weaving requires an odd number of warp threads, whereas in many forms of braiding, an even number is necessary.

6-1. *Sculpture Form,* bracelet in sterling-silver wire, Wann-Hong Liu. The techniques of winding are used to construct the initial elements which are then braided in three strands and coiled at the ends.

6-2. Necklace in sterling-silver wire, Wann-Hong Liu. Combination of techniques includes winding, rewinding, twisting, and braiding in three strands.

In braiding, no matter what the name, all of the elements are active and perform the same function in sequence, although the direction of the sequence has a number of possible variations. The elements may be worked across the entire structure from one edge only, alternately from each edge, from both edges toward the center, or from the center out to both edges. Braiding does not have to be flat. Dimensional braids in either round or square form can be produced, and in metal they have a structural sturdiness that makes them very useful elements in jewelry. Color variations can be achieved by braiding colored strands of metal in varying sequences to produce pronounced diagonal patterns of different types, such as twills, chevrons, diamonds, and lightning zigzags.

6-3. Necklace in 26-gauge wire of fine-silver, copper, and red coated copper. Eight strands of each color are used, to make a 24-strand structure approximately ⅝ inch-wide, in which each element is moved from one side over and under all the rest. The curved form is achieved during the interlacing process by pulling the strands tighter on the inside edge, and creates a permanent shape. Finishing includes weaving the ends back into the structure, coiling the silver wires at one end, and mounting three of them with pearls.

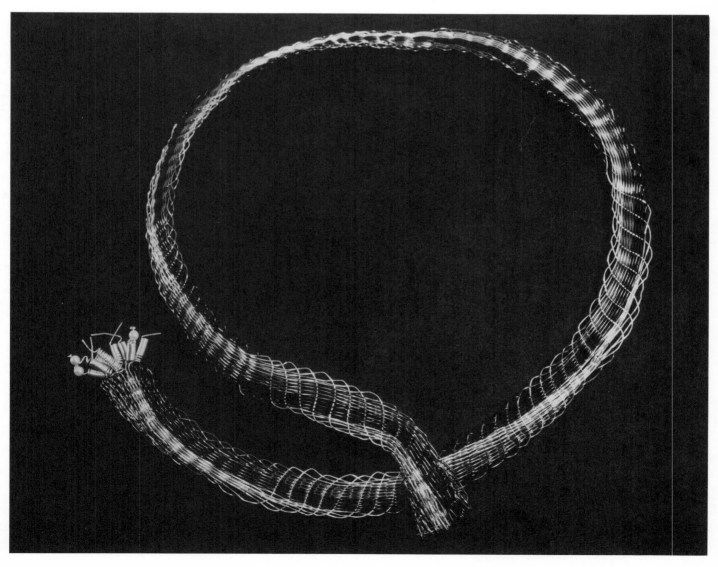

6-4. Making an eight-strand braid. Starting on the left each element is interlaced over and under across the total width. In practice, the braid will hang straight down, unless the turns of the elements are intentionally kept perpendicular to the general direction of the braid. This kind of braiding can also be done as finger weaving: alternate strands are raised and lowered, and the left-hand strand is pulled through the shed as weft. The shed is reversed, and the new left-hand strand is used as weft.

FLAT BRAIDS

The simplest form of flat braiding is the common three-strand braid. Its very simplicity makes it a decorative structure in metal (6-1 and 6-2), and it can be used as a single element or for convenient endings to pieces in other textile techniques. Three-strand braids are worked from one side only, with the outside strand being brought over and under the other two strands. It may appear that the braid is being worked in from both sides since this is the common motion used in teaching braiding to children—however, the actual motion is always from left to right (or the reverse, as long as the motion is consistent). Making a three-strand braid by moving the elements from one side only is good practice before starting on a braid with a greater number of elements.

Braiding from one side only can be done with any number of elements (6-4), and complex patterns can be created by mixtures of colors in the strands. For example, in an eight-strand braid with four strands of one color on the left and four strands of another color on the right, a diagonal stripe will be achieved. If the colors are alternated, a pattern of alternating horizontal or oblique stripes will be formed, depending on how near the horizontal you keep the strand which is being interlaced. Curved forms can be created in wire by pulling the strands tighter on the inside edge (6-3), and even compound curves can be created by controlling the tension of the interlacing elements.

Braiding from the center outward creates a chevron or herringbone pattern (6-5, 6-6, 6-7). An even number of strands is used and divided into two equal groups. Then alternately, the inner strand of each group is worked over and under toward the other side. Many interesting color variations can be achieved with this pattern, which is especially clear when several of the outer strands on each side are of a contrasting color. Rigid vertical elements at both edges, around which the wires are worked (6-6), help maintain a consistent width and pattern appearance, as well as providing a structural support. Flat wires are effective in chevron patterns, but they must be twisted back on themselves at the edges in order to maintain a consistent edge and surface. If they are allowed to twist as the work proceeds interesting surface effects and plays of light will be created. Round wires in multiple strands can also be very effective as shown in a contemporary brooch that has an almost Celtic appearance (6-7).

6-5. An eight-strand *soutache* braid of flat wire is laced around two vertical elements to establish and maintain a constant width. This pattern is achieved by lacing from the center outwards (see 6-6).

6-6. An eighteen-strand braid in three colors of flat wire. For this pattern an even number of strands must be used. Divide them into two equal groups. Then take the inner strand of the left-hand group and weave it over and under the strands of the righthand group. Do the same with the inner strand of the right-hand group, weaving it over and under the left-hand group. Continue this way, each time taking the inner strands of the groups in succession. If several of the outer strands are in a contrasting color, the chevron pattern can be seen more clearly.

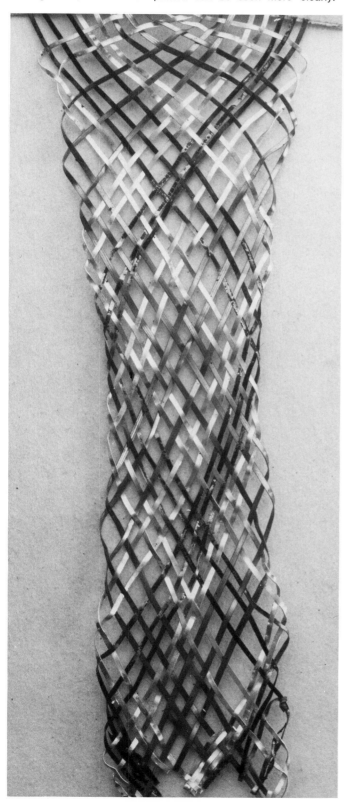

6-7. A small commercially made brooch in nickel silver uses multiple strands of round wire braided in a chevron pattern.

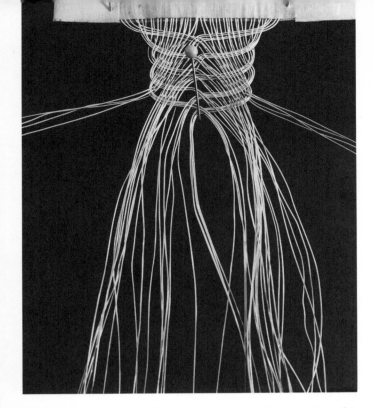

Braiding from the outer edges inward creates a different, looser appearance, with the elements moving to the opposite side at a slower rate (6-8). This sort of braid is easily separated into two parts if the inner elements do not cross each other, and the separation can be used to advantage in the construction of flowing forms which divide, change in direction, reunite, and change in width. The strands should be long enough to complete the entire form without any addition, and may even be worked in a continuous freehand manner dictated by the emerging form.

Another interesting variation, which can be achieved with greater ease in metal than yarn, is the so called Peruvian zigzag braid (6-9). This refers to the dramatic shape of the braid itself, in which the reverse turns produce sharp points at the edges of the structure, rather than the pattern within it. An uneven number of strands is worked from one side over the center strand, which remains perfectly straight throughout the entire structure, and then from the other. This braid is unusual in that it retains the vertical-horizontal interlacing pattern normally associated with weaving. It is particularly effective when multiple strips of braid are fastened together at the point or sewn together in an interlocked arrangement to form a larger structure.

6-8a. Multiple-strand braid interlaced from both sides into the center. There are 48 strands of 26-gauge fine-silver wire. There must be a cross of the two outer elements at the center, otherwise the braid will separate into two parts. This cross is achieved only after both sides have been interlaced from the outer edges to the center.

6-8b. To create a braided pendant, the large central braid has been subdivided into narrower braids which curve and reunite to produce a complex and free-flowing form. The totally continuous structure was developed in a somewhat freehand manner. Changes in size and direction were dictated by the emerging form and were easily accomplished by keeping the piece pinned to the board throughout its construction.

6-9. A "Peruvian zigzag" braid used an uneven number of strands and is worked back and forth over the center strand, which remains perfectly straight throughout the entire structure. If the center strand is in a different color the pattern is easier to see. The outer strands are interlaced first from the righthand side until all four strands are on the opposite side of the center strand. Then the strand on the far left is laced over and under to the right side, followed by all other strands in proper sequence, until all four strands again stand on the opposite side of the center strand.

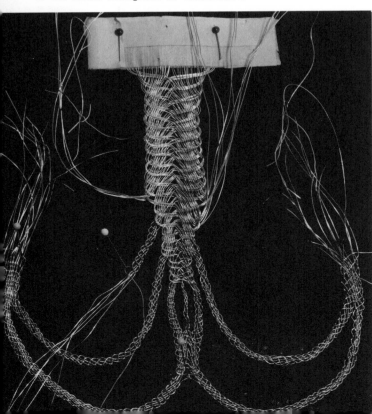

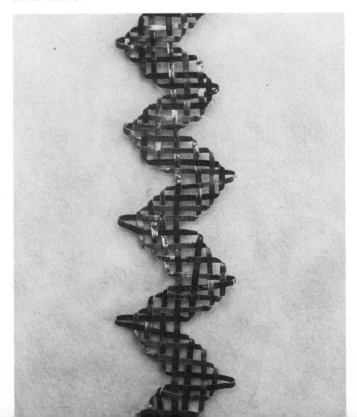

DIMENSIONAL BRAIDS

Braids can be made in round or square shapes by interlacing a number of elements divisible by four, that is eight, twelve, sixteen, twenty elements, etc. With wire, dimensional braids are generally self-supporting and extremely durable because of their density. The elements are constantly crossing over each other in the center of the braid making the end result an almost solid structure. Depending on the original arrangement of different colored elements, and the sequence in which they are used, quite intricate patterns of spirals, chevrons, zigzags and diamond shapes can be developed.

The braid can be made around a core in order to increase the diameter without increasing the number of elements needed. When braiding in yarn, the core is usually a heavy cord or a large mass of yarns twisted to form a core. It is possible to braid thin strips of metal around a similar cord, but the resulting surface is not as even as when braiding around a heavy wire or metal rod. Working the structure around a core greatly simplifies the actual interlacing movement by making the pattern larger and easier to follow without error in sequence. It also enlarges the pattern significantly to give a more decorative effect. The metal core can be easily removed if it has been coated with a lubricant like oil or petroleum jelly or can remain if the additional weight and support are desirable.

Both round and flat wires can be used with equal ease, providing the flat wires are not allowed to twist as they are being worked. For braids of any reasonable dimension, that is, over one-quarter inch in diameter, flat wire should definitely be used instead of round since it produces both a larger and more compact structure. Round wires must ride over each other the distance of at least twice the diameter of the wire and are difficult to pull tightly together so the braid appears to be loosely done and the pattern is difficult to distinguish clearly.

In the round braid in 6-10 the total diameter is three-sixteenths of an inch, since it is constructed over a one-eighth-inch diameter rod. The same braid done over a larger core with the same wire would produce an equally compact surface but the pattern would be horizontally compressed. For more dramatic changes, the scale or the number of elements used should be increased as the diameter of the core is enlarged. As mentioned, the number of elements must be in multiples of four; in this braid the outer element goes under all of the elements of one side and half of the elements on the other before reversing direction.

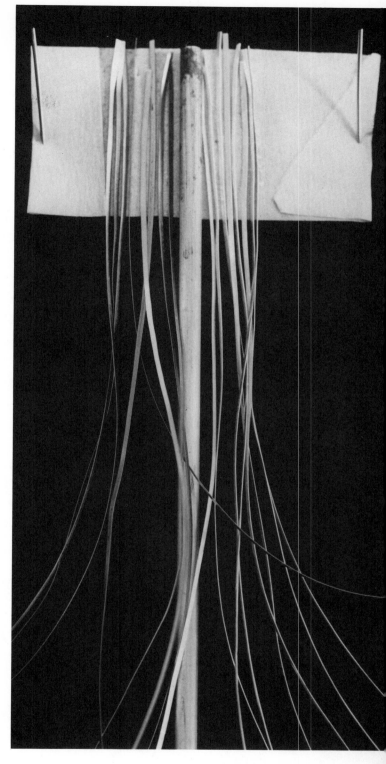

6-10a. Round braid, step one. Sixteen strands of flat wire are arranged equally on both sides of a central rod to begin a round braid. A brass rod is used here with copper elements on the left and silver elements on the right. The elements are held in place by strips of masking tape.

6-10b. Round braid, step two. The chevron pattern is created by bringing the outer left strand under all the left strands, the center rod, and half (in this case four) of the right strands. It then reverses direction, as shown, and is brought over and across the right strands and center rod and placed directly to the left of the rod. The same motions are made with the outer element on the opposite side.

6-10c. Round braid, completed. The central core can be removed without any damage to the round cross section of the braid.

The simple four-strand braid in 6-12 uses the same technique as lanyards made in plastic gimp or dog leashes in leather, and is best made over a central core in order to increase its diameter. This braid is so compact that the core can be removed, and the remaining hole used for stringing the elements as beads (6-11). As with the other techniques discussed, the function and visual impact can be changed drastically by changing the scale of the material and the finished form. The same process that produces the one-half-inch beads shown here could make a free standing, self-supporting sculptural form of almost any height and diameter.

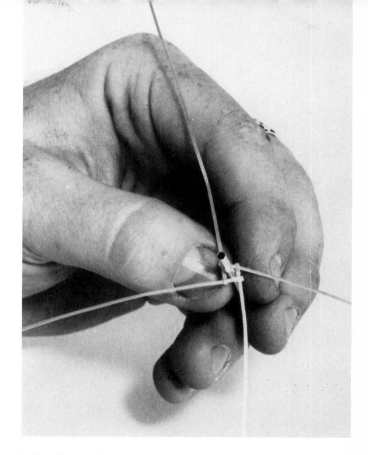

6-12a. Square braid, step one. The four strands of cloisonné wire in 30-gauge, $\frac{1}{16}$-inch-width silver wire are locked over each other and arranged around a round central tube or rod. The first layer is made by folding each wire over on itself and interlacing it with the next wire in perpendicular rotation. The last wire is slipped through the fold of the first and all four ends are pulled to tighten the structure.

6-11. Necklace of square-braid beads, Steven Brixner. The beads are 30-gauge fine-silver cloisonné wire, $\frac{1}{16}$-inch wide.

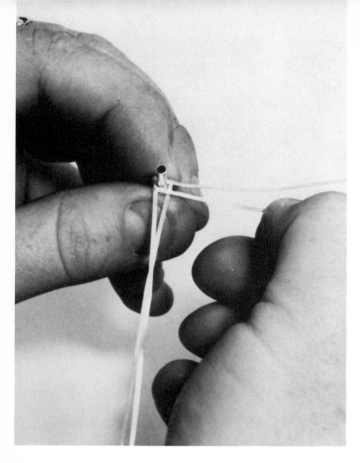

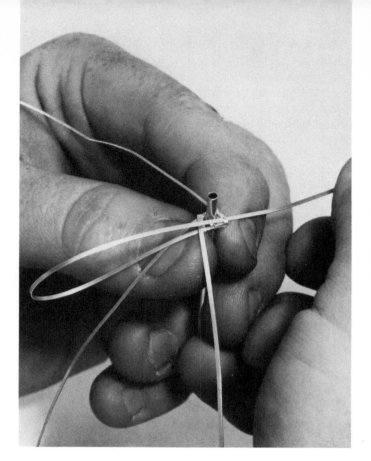

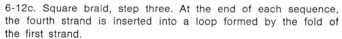

6-12b. Square braid, step two. The strands are folded over each other in rotation.

6-12d. Square braid, step four. The strands are pulled tightly together.

6-12c. Square braid, step three. At the end of each sequence, the fourth strand is inserted into a loop formed by the fold of the first strand.

6-12e. Square braid, step five. When the braid is long enough, it is finished by clipping the ends of the four wires after they have been locked in place under the other wires. This demonstration series was prepared by Steven Brixner.

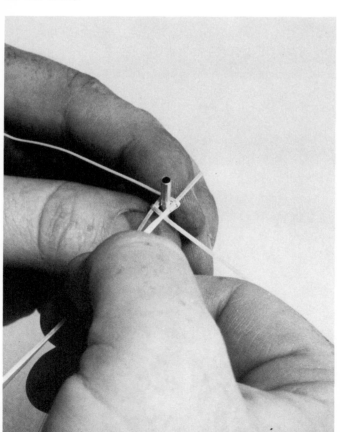

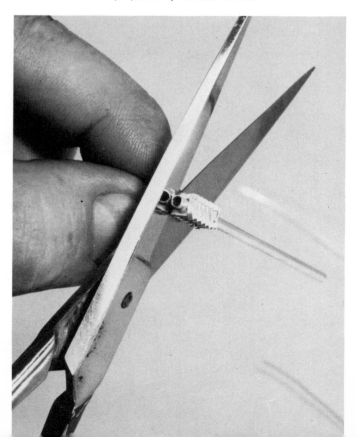

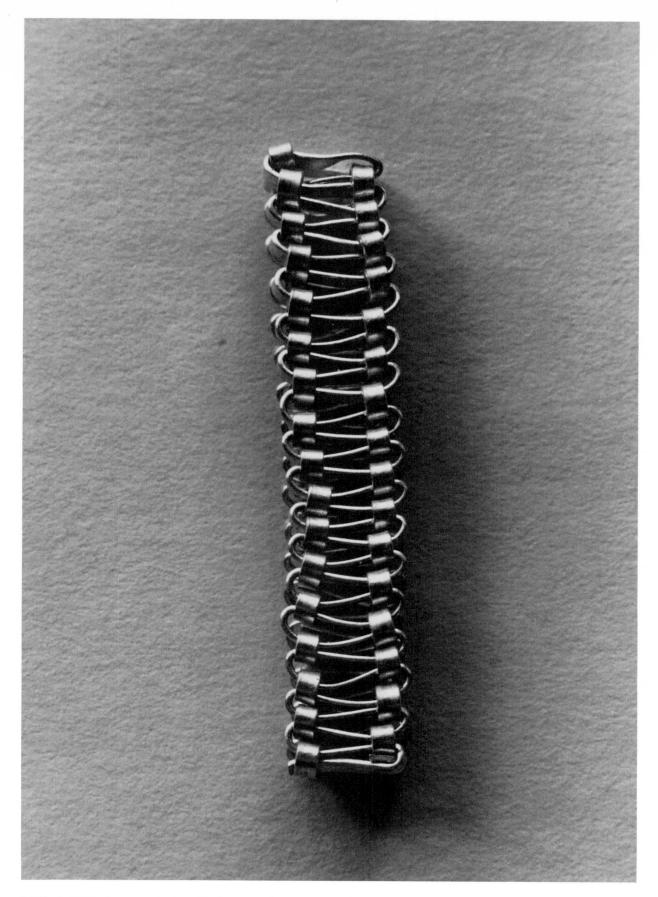

6-12f. A finished square braid with the central core removed. The resulting hole can be used for stringing the braid as a bead (see 6-11).

7/Interlinking and its Variations, Sprang and Bobbin Lace

Interlinking, like braiding, involves a single set of elements placed next to each other in a more or less parallel manner from a common starting point. All the elements operate in a similar manner to produce the structure. In braiding, the elements move all the way or halfway to the other side of the structure, but in interlinking they twist around adjacent elements on both sides, either in a regularly alternating zigzag pattern or in any other combination that allows the elements to move in a generally vertical pattern, changing their relative positions only slightly in the process.

Bobbin lace is closest to braiding in structure, but uses multiple elements that do not all link with one another as they do in a braid. A braid forms a compact structure because of its constant interlacing, whereas the structure of bobbin lace is very open. Sprang is similar to basic interlinking but the elements are attached and under tension at both ends, rather than just at the top of the structure.

A simple example of the interlinking process can be observed in the ordinary chain-link fence (7-1). Multiple elements are arranged in pairs at the upper edge but operate as evenly spaced single elements. Each element moves either to the right or left, twisting around the adjacent element, and returns to its original position. This process creates a continuous series of diamond shapes, which are structurally rather rigid because of the stiffness of the wire used. In all interlinking the basic integrity of the form is retained as the result of the constantly alternating zigzag path of the elements. No additional locking device such as a weft is needed for the structure to be totally self-sustaining. Since it is only a two-dimensional plane it is not self-supporting without an additional frame.

One variation of the basic interlinking structure is accomplished by additional twists in the same link between two elements (7-2). Double, triple, or even quadruple twists produce elongated diamond-shaped spaces in a more open structure because of the larger area between the links. Another variation is achieved by moving a single element in a less regular path than the simple zigzag of the chain-link fence. The element may move to the right for two or more twists before returning to its original position, or it may continue to one side, never returning, to produce a slightly spiral effect.

More complex patterns can be created by making holes or spaces in the regular pattern (7-4). The unused elements are carried along the perimeter of the open spaces and have the effect of outlining the holes and emphasizing the open-work pattern. The color effects that can be achieved by combining different color sequences in the original set-up of elements, as well as pattern variations in the interlinking, can produce an unlimited number of intricate lacelike patterns which would be applicable to both delicate jewelry and large-scale screens and wall hangings.

The necklace in 7-2a is a good example of interlinking. Although the result looks very similar to sprang, the open curved form of the necklace could only have been achieved by working down from the top. A crocheted chain was made first, and then a double length of sterling silver wire was inserted into each chain loop to form one pair of elements, making a total of 60 pairs. Each row of twists was done in horizontal sequence, and each twist held in place by a pin. The subsequent row locked the twists in place so the pins could be removed as the fabric developed. The twists are alternated from row to row, being worked right over left and then left over right, which produces a particular pattern and keeps the elements in their original sequence as they move through the width of the band. The necklace was finished at the bottom with another length of crocheted chain stitch held in place by an additional three twists. This ending confines the expansion of the bottom edge to conform to the necessary shape for a fitted collar. The basic shape of the collar form is critical to a proper fit and the pattern for the piece must

7-1. A section of ordinary chain-link fencing is a simple example of interlinking. Each of the multiple original elements twists around the adjacent element to return to its original position, creating a series of diamond shapes.

be determined in paper or muslin prior to working in metal. The necklace is fastened in the back with a series of hooks extending for the entire width of the piece. There is sufficient "give" in the interlinked structure to allow for a constant reshaping of the finished form without damage, providing only the hands are used and no sudden bends or folds are made.

As these examples show, interlinking lends itself well to work in metal because it causes no kinking or stress and very little work hardening in the metal. Interlinking can be used with metal of any scale since no particular mechanism or instrument is required and the only movement is a simple twist or cross-over between pairs of elements. Bobbin lace is also very easy on metal, and with improvised bobbins could be done with very large-gauge metal wire as long as the material was malleable enough to retain the pattern. Sprang, however, is worked on a warp attached at both ends, so that the interlinking process causes great stress on the individual elements. For this technique, then, quite flexible and strong wires are a good choice, as they are not so likely to snap from the tension.

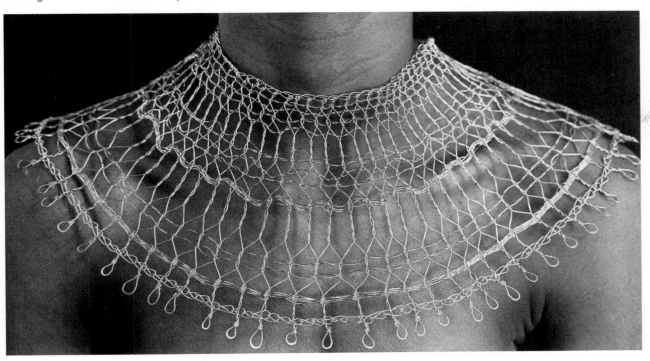

7-2a. Interlinked necklace in sterling silver wire of 18-gauge. Begun with a crocheted chain and worked flat, this necklace is constructed by twisting the multiple elements around each other in a pattern that depends for its effect on variations in the number of twists in each pair of elements. The horizontal elements are decorative, not structural. This technique is closely related to sprang (compare 7-2b and 7-5) and the twists were held in place with pins as work progressed, a method which is used in bobbin lace (see 7-11b).

7-2b. Detail of necklace showing the alternation of twists (from top): single, single, triple, single, quadruple, double. The spacing was gradually expanded to produce a curved form that can be fitted around the shoulders as well as being returned to its flat position without damage to the structure.

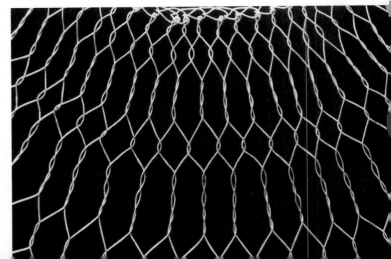

SPRANG

Sprang is a variation of interlinking in which the set of elements is stretched between two beams or across two ends of a rigid frame. As the elements are twisted around each other, the interlinking occurs simultaneously at both ends, producing a symmetrical structure at top and bottom. The work is generally done from the bottom up—and automatically also forms from the top down in a mirror image—terminating in the center, which is the only place it has to be permanently locked in place.

Sprang tends to be associated with Scandinavia, where it has been used from Viking times until the present. It occurs in many other areas historically, including ancient Greece, Luristan, Syria, Coptic Egypt, and pre-Colombian Peru, and still survives in both North Africa and among the Indians of North America. The bisymmetry of the process has often been used to advantage in these cultures in producing shaped garments such as caps and hats. The natural effect of the technique is usually a marked compression of the form as it moves toward the center (see 7-4), and the indentation is often employed structurally in making the peak of a cap or the center seam of a shaped hat.

The same pattern possibilities are available in sprang as in simple interlinking. Zigzags, multiple twists, and shaped openings can all be combined to produce intricate lacelike patterns. The only real difference in the end product is the mirror image that results as a consequence of the attachment of the warp at both ends instead of only one. Even the twists are reversed in direction. If a z-twist was made at the bottom an s-twist forms at the top. This change is barely discernible and does not make a difference in the pattern or shape, although in very complex patterns account must be taken of which direction was used to form the last row of twists.

Using the sprang technique in metal requires that at least one of the supporting beams be moveable to compensate for the lack of stretch quality in the material. The tension is very great, even in yarn, because two twists are being made at each step, not one. The elements are in effect being shortened, and must move closer to each other periodically to reduce the tension and the center compression. This is easily accomplished if one beam is tied in position, with a means for gradually loosening the ties as the work progresses, a system that is sometimes used in weaving. The more the tension is relieved, the easier it is to twist the elements and to avoid a severe compression in the center of the piece when it is not desired.

It is possible to avoid the center compression completely by using side supports to pull the structure horizontally. In yarn, these are merely additional cords tied through the vertical edges at frequent intervals to the frame. They must remain to control the shape in a yarn piece, but in wire the structure will be sufficiently sturdy so that it can maintain whatever shape is made with all the temporary supports removed.

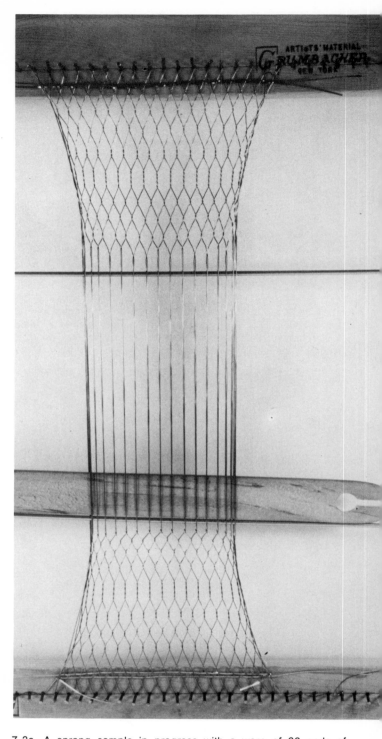

7-3a. A sprang sample in progress with a warp of 36 ends of 28-gauge copper wire strung on a simple frame (an even number of warp threads must be used). The mirror-image of the pattern being worked at the bottom can be seen clearly at the top. After each row of twists is made with the stick, the two knitting needles are inserted in the shed and pushed to the top and bottom to hold that row of twists in place. When the next row is made, the knitting needles are removed from the last row, and again inserted in the shed.

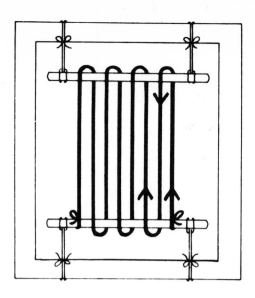

7-3b. Sprang can be done on a simple frame.

The only instruments used in sprang, beside the frame, are two rods or sticks for holding the links in place until they are locked in by the next rows, and a wide stick for picking up and twisting the adjacent elements in horizontal order. This stick can be turned on edge to form the shed into which the holding rods are placed, and then pushed to the top or bottom until the next row is formed. All sprang must be warped with an even number of threads, and the rows are made in a two-row sequence (7-3b,c,d,e). If the sequence is not followed, and all rows are made in exactly the same way, only long ropelike forms are created, rather than the characteristic diamond zigzags. Openings are created by variations on one of the rows in this sequence, while the regular alternation with the other row continues. They are closed by going back to the regular sprang sequence (7-4).

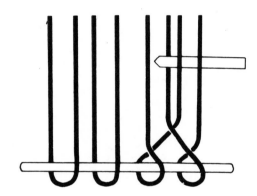

7-3c. Sprang, step one. The sequence for sprang must be an alternation of two kinds of rows. In the first, two strands are twisted over one strand to begin.

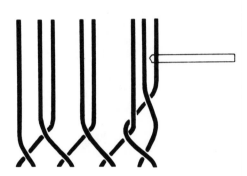

7-3e. Sprang, step three. The second row begins with a simple twist.

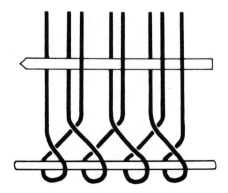

7-3d. Sprang, step two. The row is continued with simple twists until the last, in which one strand is twisted over two strands to end.

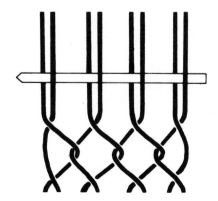

7-3f. Sprang, step four. The second row contains simple twists all the way through. This two row sequence is basic to all sprang designs, no matter how complex they become. Drawings by Lisa Ridenour, from *Sprang*, by Hella Skowronski and Mary Reddy.

When the center is reached and no further twists can be made, some holding element must be inserted to lock the entire structure. This may be a piece of wire or metal woven in as a decorative element or it may be a rod of some other material which can remain in the structure permanently. A less obvious device is to make a series of chain stitches with a crochet hook. The hook picks up each of the elements in succession from left to right (7-5). This produces a decorative effect without the addition of a supplementary element.

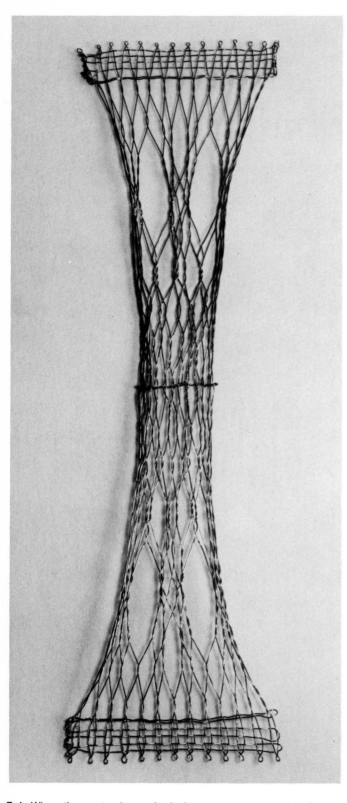

7-4. When the center is reached, the structure must be locked in place. A wire might be threaded through the last shed; here a double strand of wire is twined around each warp element. This sample, in 26-gauge copper wire, uses complex twists to produce a pattern of openings. Each group of twists before and after the opening is spranged like a complete first row in the basic sprang sequence (7-3c). The height of the opening is determined by how many pattern rows are made, each row alternating with a simple second row (7-3e). The opening is closed by spranging the first row (7-3c) across the whole warp.

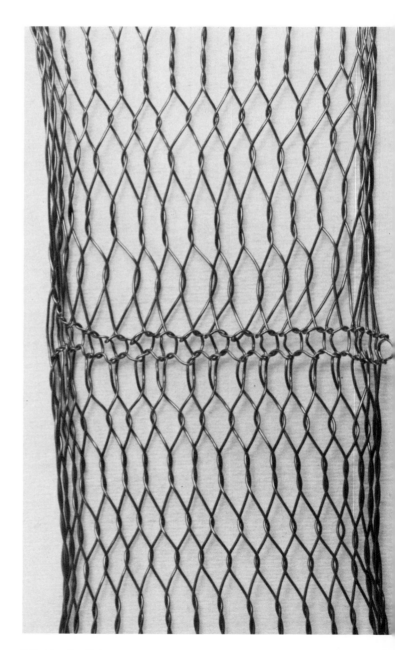

7-5. Detail of the center of a sprang construction finished with a series of interlocking chain stitches made with a crochet hook. Single, double, and triple twists can be seen here.

124

BOBBIN LACE

The making of bobbin lace involves an elaborate form of interlinking of a single set of elements starting from a common point to produce complex and highly refined patterns. The specific process is thought to have originated in Germany in the sixteenth century as a way of controlling large numbers of elements of unrestricted lengths, although there is every evidence that similar structures were produced in much earlier cultures, probably using implements not so very different from the bobbins and pillow used today. Other types of laces are generally made with a single thread and a needle and are called by a variety of names: needlepoint lace, crochet lace, embroidered lace, and knitted lace.

Bobbin, or pillow, lace refers specifically to those structures which are comprised of multiple elements, each element being wound on a small bobbin of wood or bone to give weight and to allow for greater length. The elements are interlinked in a sequence dictated by the specific pattern. The structure is formed on a "pillow," which may be any kind of raised platform or roll which allows pins to be inserted at the twists and crossings of the elements to hold them in place until the next sequence of rows locks them permanently.

The implements used for this technique have been refined over several centuries to perform their function as swiftly and easily as possible. Watching the flying fingers of lacemakers in Portugal, Sicily, or Malta, as they manipulate hundreds of bobbins wound with the finest threads to produce their beautiful and highly prized laces, one is aware of the absolute refinement of tools which literally slip through the hands like water. The bobbin is a small elongated spool with a slight swelling at the end to form a kind of handle. Most of them are wood, lathe-turned in a great variety of shapes, although one can sometimes find handsomely crafted ivory or bone versions. The contemporary bobbins are rather clumsy and large scale in comparison with traditional ones, and have an unpleasant feel of raw wood; most of mine were collected in antique and junk shops, giving me a lovely assortment of shapes. The variety is both pleasing to the eye and easier to work with when a large number are being used at one time.

The bobbins are wound either by hand or with a lacemaker's or weaver's mechanical bobbin winder with as much material as they can hold comfortably. Yarn must be knotted at the top of the bobbin with a half-hitch to keep from unraveling but this is not necessary when using wire. A "pillow" or platform (see 7-11a) is essential for two reasons: it elevates the work area, allowing the bobbins to hang out of the way when not being used, and it provides a soft ground for holding the pins that support the work as it progresses. The most usual kind of pillow is a cylindrical roll of stuffing wound over a wooden rod which has large wooden discs at either end. The stuffing is covered with a tight-fitting sleeve of closely woven

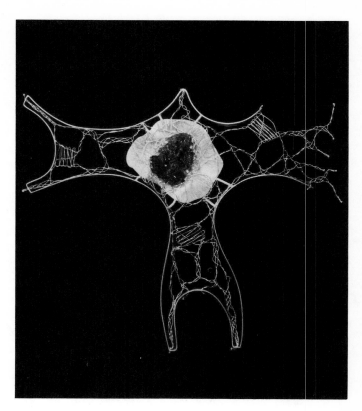

7-6. *Lacy Web*, brooch of 18-karat gold with a Chatham ruby crystal. A contemporary adaptation of bobbin lace was developed in a radial pattern while pinned to a soft board. The ends are inserted into holes in a sheet-metal frame, and are melted into beads which hold them into position permanently. The close relation to sprang and braiding can be seen in this simple bobbin-lace structure.

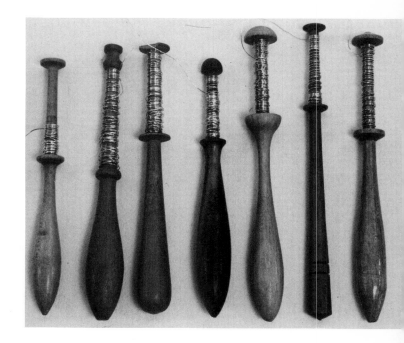

7-7. Antique lace bobbins of wood. These are wound with thin gauges of fine-silver wire of 26 and 30 gauge). The winding, which should be done as evenly as possible, and always in a clockwise direction, may be done with the aid of a bobbin winder. When winding bobbins with thread, a slip knot is made at the top end to help control the unwinding of the thread but with metal the coils are tight enough so this is not necessary.

fabric, felt, or suede cloth. The roll must be wide enough to accommodate the pattern and large enough in diameter to allow for a sizable section of work to be completed before turning. A good average size would be 8 to 10 inches in width with a 4 to 5 inch diameter. The roll is supported by slipping the ends of the rod into upright fixtures fastened to a table or stand so that it can be turned as the work progresses. A simple platform of soft mill-board placed on a turntable is also quite adequate, especially if the finished piece is small in size and the elements are moved in a radial pattern. The only other equipment needed is a box of pins, and long dressmaker pins (with round glass heads) are the most practical since they hold the crossing more firmly and are easier on the fingers than the ordinary straight pins.

A simple set of movements involving two pairs of bobbins produces the *lace stitch,* which is the basic stitch in all patterns, even the most complex. The four bobbins are moved in two different ways. First, the right-hand bobbins of each pair are *twisted,* which means that they are placed over the left-hand bobbins of their own pair (7-8b); then the inner two bobbins are *crossed,* which means that the left-hand one of the inner pair is placed over the right-hand one, leaving the two outer bobbins in place (7-8c). At this point a *half stitch* has been made; when these two sets of movements have been repeated once, and the pairs are reunited, a *whole stitch,* or *lace stitch,* has been completed (7-8d,e). Since the movements which have been completed are twist, cross, twist, cross, they are often designated on pattern diagrams as TCTC (see Appendix for list of abbreviations).

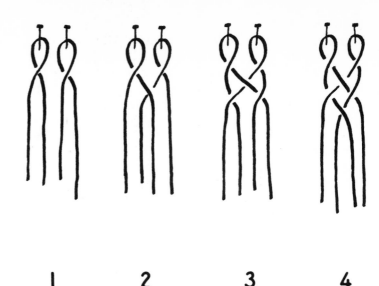

1   2   3   4

7-8a. Basic stitch in bobbin lace. Two pair of bobbins are needed for the basic stitch. They are first crossed right over left (1), then the center pair is crossed left over right (2), then the two pairs are crossed right over left again (3), and finally the center pair is crossed left over right again (4). The whole sequence, designated as *twist, cross, twist, cross* forms the *whole,* or *double, stitch* also sometimes called the *lace stitch.* The first two steps, *twist, cross,* form a *half stitch,* and the first three steps, *twist, cross, twist,* form a *linen stitch.* Sometimes lace makers use the sequence in the reverse order—*cross, twist, cross, twist*—but this does not alter the structure essentially.

7-8b. Bobbin lace, step one. The right bobbin in each pair is placed across the left bobbin (*twist*).

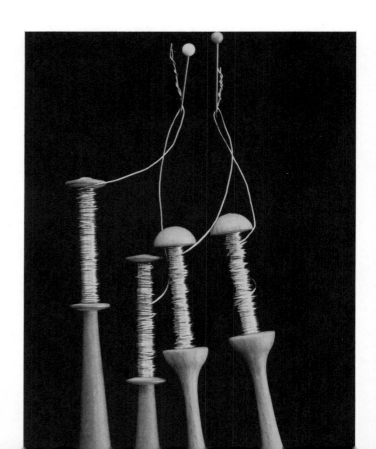

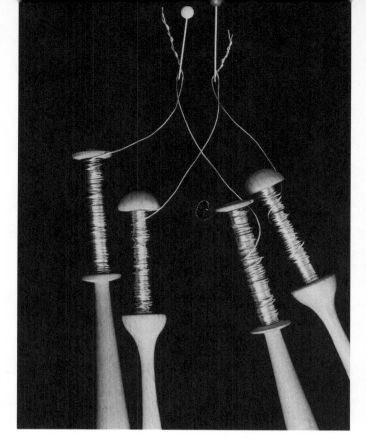

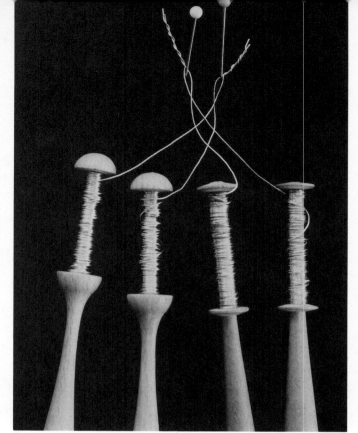

7-8c. Bobbin lace, step two. The left bobbin of the center pair is placed across the right bobbin (*cross*).

7-8e. Bobbin lace, step four. The left bobbin of the center pair is placed across the right bobbin (*cross*). This completes one double stitch; the pairs of bobbins are reunited but on the opposite side from where they started.

7-8d. Bobbin lace, step three. The right bobbin in each pair is placed across the left bobbin (*twist*).

7-8f. Bobbin lace, step five. The pattern can be continued as long as required. With only a single pair, the series of whole stitches resembles a four-strand braid. All bobbin lace structures use many more than two pairs of bobbins (see 7-11a).

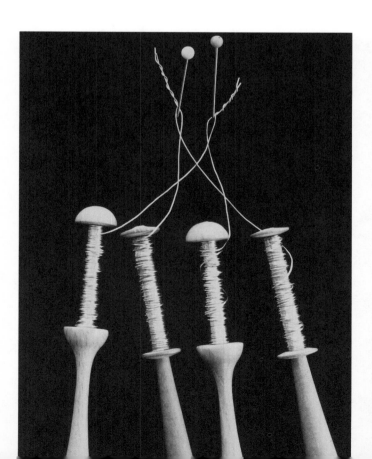

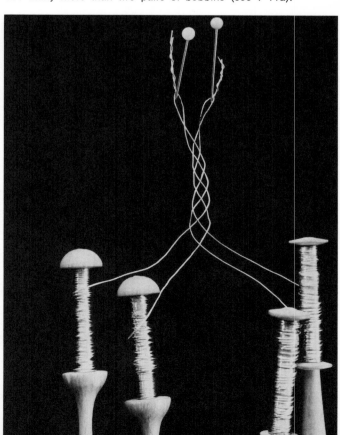

Most lace patterns involve a certain amount of ground area which is really a netted fabric composed of various pattern combinations of lace stitch within which more elaborate figures can be developed, either with the same bobbins or with supplementary or inlaid elements. A simple exercise in making a ground pattern is invaluable in developing some skill in using the basic movements as well as in understanding the diagonal sequences which constitute the structural development. Of the many grounds developed by bobbin lacemakers, two basic ones, Torchon (7-9) and Paris (7-10) are presented here in diagram form. Each of the numbers indicates a place in the structure where a pin is inserted. The instructions for the pattern sequences follows. In both cases a band approximately one inch in width is produced, using 30-gauge fine-silver wire. The patterns can be repeated an infinite number of times to make a band of any length, which could be incorporated into jewelry such as bracelets or used to embellish clothing.

Instructions for Torchon Ground

Hang 2 pairs of bobbins at
A, B, C, D, and E.

2 &	3	pr. TC	Pin at	1	TC
1 &	2	pr. TCTC	Pin at	2	TCTC
4 &	5	pr. TC	Pin at	3	TC
3 &	4	pr. TC	Pin at	4	TC
2 &	3	pr. TC	Pin at	5	TC
1 &	2	pr. TCTC	Pin at	6	TCTC
6 &	7	pr. TC	Pin at	7	TC
5 &	6	pr. TC	Pin at	8	TC
4 &	5	pr. TC	Pin at	9	TC
3 &	4	pr. TC	Pin at	10	TC
2 &	3	pr. TC	Pin at	11	TC
1 &	2	pr. TCTC	Pin at	12	TCTC
8 &	9	pr. TC	Pin at	13	TC
7 &	8	pr. TC	Pin at	14	TC
6 &	7	pr. TC	Pin at	15	TC
5 &	6	pr. TC	Pin at	16	TC
4 &	5	pr. TC	Pin at	17	TC
3 &	4	pr. TC	Pin at	18	TC
2 &	3	pr. TC	Pin at	19	TC
2 &	1	pr. TCTC	Pin at	20	TCTC
*9 &	10	pr. TCTC	Pin at	21	TCTC
8 &	9	pr. TC	Pin at	22	TC
7 &	8	pr. TC	Pin at	23	TC
6 &	7	pr. TC	Pin at	24	TC
5 &	6	pr. TC	Pin at	25	TC
4 &	5	pr. TC	Pin at	26	TC
3 &	4	pr. TC	Pin at	27	TC
2 &	3	pr. TC	Pin at	28	TC
1 &	2	pr. TCTC	Pin at	29	TCTC

Repeat from *

Instructions for Paris Ground

Hang 3 pairs of bobbins at A & C
Hang 4 pairs of bobbins at B

4 & 5	pr. TCTC		
2 & 3	pr. TCTC		
3 & 4	pr. TC	Pin at	1 TC
2 & 3	pr. TCTC		
1	pr. T alone		
1 & 2	pr. TCTC	Pin at 2 between 2 & 3 pr.	
2 & 3	pr. TCTC		
8 & 9	pr. TCTC		
6 & 7	pr. TCTC		
7 & 8	pr. TC	Pin at	3 TC
6 & 7	pr. TCTC		
4 & 5	pr. TCTC		
5 & 6	pr. TC	Pin at	4 TC
4 & 5	pr. TCTC		
3 & 4	pr. TC	Pin at	5 TC
2 & 3	pr. TCTC		
1	pr. T alone		
1 & 2	pr. TCTC	Pin at 6 between 2 & 3 pr.	
* 8 & 9	pr. TCTC		
10	pr. T alone		
9 & 10	pr. TCTC	Pin at 7 between 8 & 9 pr.	
8 & 9	pr. TCTC		
6 & 7	pr. TCTC		
7 & 8	pr. TC	Pin at	8 TC
6 & 7	pr. TCTC		
4 & 5	pr. TCTC		
5 & 6	pr. TC	Pin at	9 TC
4 & 5	pr. TCTC		
3 & 4	pr. TC	Pin at	10 TC
2 & 3	pr. TCTC		
1	pr. T alone		
1 & 2	pr. TCTC	Pin at 11 between 2 & 3 pr.	

Repeat from *

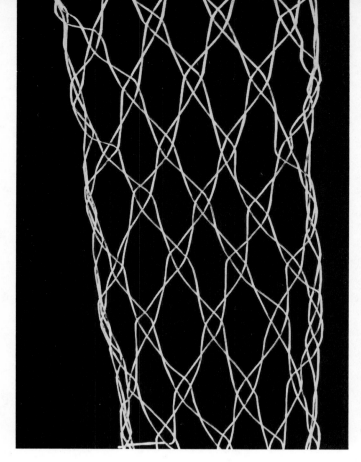

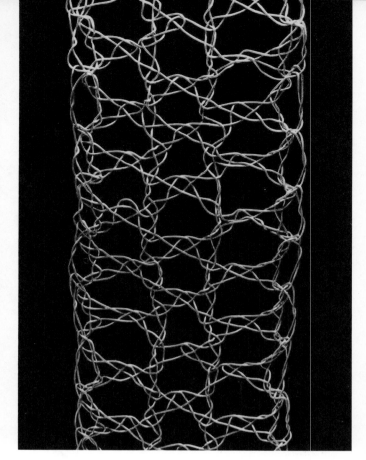

7-9a. A simple Torchon ground pattern for bobbin lace. This sample is done in 30-gauge fine-silver wire wound on ten pairs of bobbins, two pairs hung from each of five evenly spaced pins, and produces a band approximately 1 inch wide.

7-10a. A sample band of Paris ground also in 30-gauge fine-silver wire on ten pairs of bobbins. Three pairs hang at each of the outer edges and four pairs hang from the center pin.

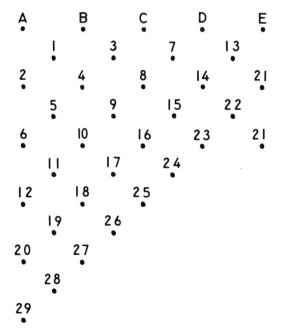

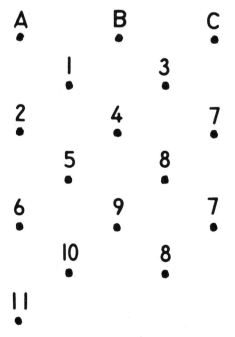

7-9b. The pattern for a Torchon ground. Pattern papers, sometimes already pricked with holes, are placed on the bobbin lace "pillow" and the lace is made over them. Each of the numbers indicates the placement of a pin after the required stitches have been made. The instructions for Torchon ground may be found in the text.

7-10b. The pattern for a Paris ground. The instructions may be found in the text.

A more complex pattern called Labyrinth, also in 30-gauge fine-silver wire, uses a supplementary pair of bobbins wound with 26-gauge fine-silver wire. This extra element is inlaid within the ground to form a linear pattern that outlines a series of holes in the base structure. In this case only a printed pattern with directional lines was used as a guide. It was pinned around the pillow, and the wire structure was worked directly on the paper diagram. Such printed diagrams frequently accompany instructional books on lacemaking and can be used in the same size or copied at the desired scale on heavy-weight paper.

Once the basic technique is mastered it is possible to improvise any number of individual patterns. It is not really necessary to work in bands; the technique lends itself equally well to radial and omni-directional patterns. Nor is it necessary to repeat a single pattern over and over again to form a given piece. The gold brooch in 7-6 was begun on a radial scheme but gradually evolved into a meandering three-pronged form with occasional repeats of a square-woven motif placed in a random sequence. This piece is open in effect and not very sturdy structurally because so few bobbins were used. When working out from a center the incorporation of additional pairs of bobbins as the area increases would add strength.

Lace is generally thought of as a delicate technique to be used only in small scale. Certainly, if it is done on a pillow the dimension is very limited. However, the basic structure will work in any scale and with any size material provided that the relationships are kept in proper balance. The same gauge wire was used to produce the samples in 7-9a, 7-10a, and 7-11c, but the more elaborate pattern is wider because more bobbins were used. Increasing the size of the wire substantially would obviously require a larger area to be worked if any discernible pattern is to be achieved. Moving from a pillow to a platform as the working surface, also increases the freedom to work in larger patterns and less restricted shapes, as well as to move in many directions at one time. The technique of bobbin lace has great possibilities for improvisation, especially in metal, but one should follow a traditional pattern accurately first to gain mastery of the technique.

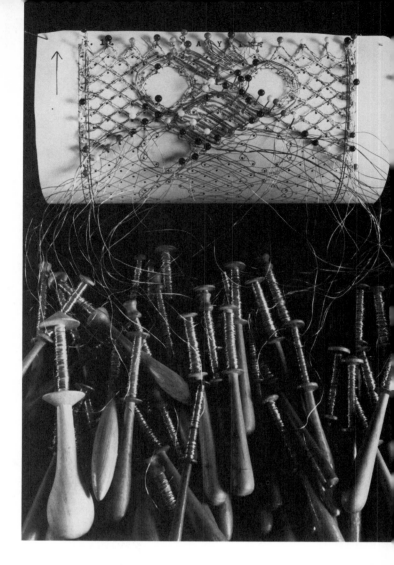

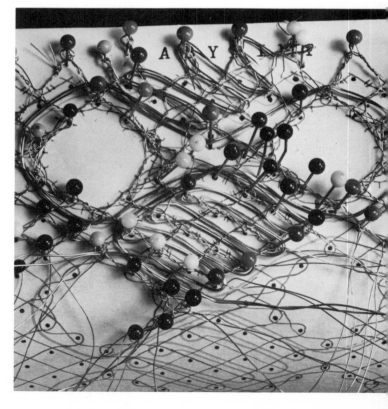

7-11a. Labyrinth pattern in progress on a cylindrical pillow. This more complex bobbin lace uses 28 pairs of bobbins wound with 30-gauge wire and two additional pairs of bobbins wound with 26-gauge wire to create a supplementary linear element inlaid into the regular ground pattern. In bobbin lace this ground pattern is sometimes called *cloth,* and is done with thin cord.

7-11b. Detail of the Labyrinth pattern, showing how the heavier-gauge wire is used to outline a series of holes in the ground structure. The printed pattern used as a guide can be clearly seen. There are no numbers, just directional lines, and the lacemaker follows the guide directly without referring to a series of written pattern directions.

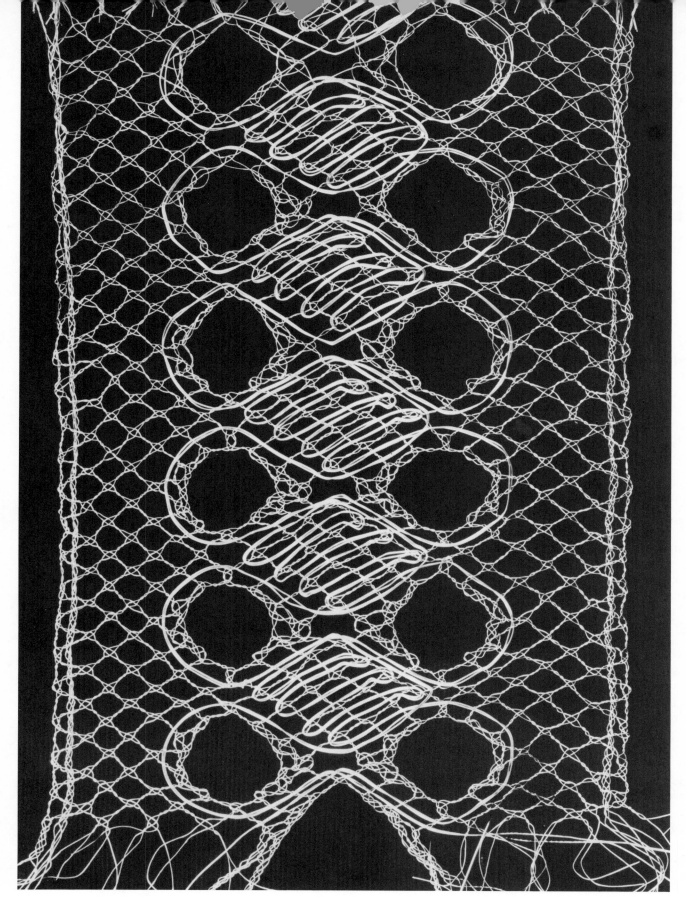

7-11c. Finished sample of Labyrinth pattern, 5 inches wide. Finishing of bobbin lace in wire is a simple task. The ends are merely twisted together and clipped for beading or other ornamental endings. Pattern courtesy of Nyrop-Larsen, *Knipling Efter Tegning*.

8/Basketry

Many of the techniques used in plaiting, twining, coiling, caning, and rushing are readily adaptable to use in metal. The materials traditionally used, such as reeds, grasses, rushes, splints, vines, and coconut and palm products, are somewhat like metal in that they are sturdy and flexible at the same time. Some of these fiber products have to be continually dampened to keep them flexible during the basketmaking process, which is not necessary with metal.

A very interesting transposition in materials from coconut palms to sheet metal can be found in a "noir" or coconut box from Perak in Malaysia (8-1). Made in silver, in a close imitation of the same sort of boxes made from coconuts, it has a lid cut in a serrated pattern just like a coconut cut with a knife. The sphere has been formed in two hemispheres and the joint hidden with a thin plaited silver band like those once made from the unopened frond of the coconut palm. Closer to home one can find examples on the commercial market of twined baskets in gilded metal, like those shown in Chapter 3 (3-4). As early as the eighteenth century bread baskets were translated into silver, although modern methods of manufacture quickly cheapened the process, and instead of being woven in wire the baskets were soon constructed of pierced sheet metal. Today one can find elegant woven fruit baskets in silver and vermeil (gilded silver, bronze, or copper) meant to be set on coffee tables to hold real or silver fruit.

With metal, a much broader interpretation of the various basketry processes is possible. The techniques can be used to form dimensional shapes for jewelry, decorative sculptures, and containers of various function and style. Few tools are needed in basketry, and generally the work can be done in metal in the same way as in fiber. Wire can be easily substituted for the lacing materials used in twining (8-2) and coiling (8-3), and flat strips of metal or flat wire can serve well in place of the splints, fronds, and canes that are used

in plaited and woven work. Beginning and finishing are, if anything, easier than with fiber materials because the metal can be fastened permanently in place by soldering rather than having to be secured by tying. There are solutions possible in metal that are not possible with fiber. For example, the spokes in a twined form can be widened at the ends by hammering (8-2b, 8-7b) so that the twining does not slip off, rather than being bent back into the basket.

8-1. A silver box from Malaysia in the form of a coconut with a plaited band of silver covering the joint where the two spheres come together.

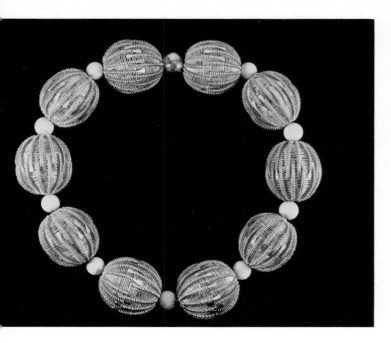

8-2a. Basket-woven beads by Steven Brixner. The beads are constructed in the technique of twining in 26-gauge fine-silver wire over a radial warp structure of 12-gauge sterling-silver round wire. The spacer beads are ivory, and the clasp is constructed from a hollow silver bead (at top).

8-3. *Neckpiece #6*, Mary Lee Hu. Individual elements of fine silver over sterling silver are constructed in the technique of coiling and then assembled into larger units by further coiling and wrapping.

8-2b. Detail of the beads. They are made in two halves which are force-fitted over one another. Warp elements are arranged radially and soldered to a center ring. When twining is finished the warp ends are widened by hammering. The beads and spacers are strung tightly on a chain.

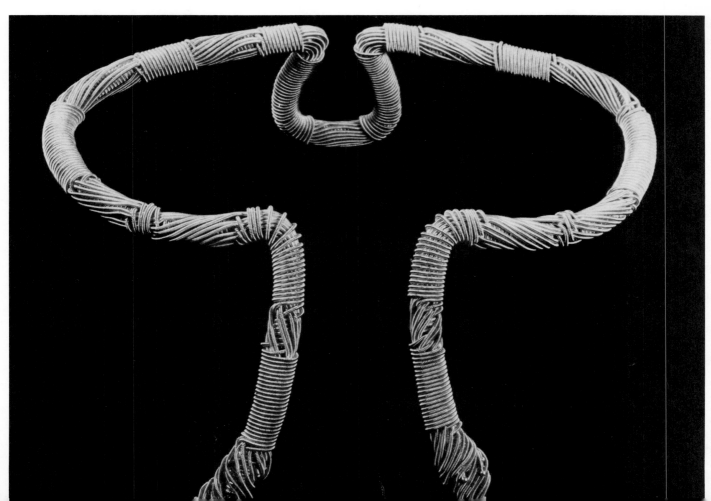

PLAITING

In basketry, plaiting is used to make a great variety of of forms, from flat mats and rectangular bags to containers of round, oval, square, and rectangular cross-sections. Plaiting in basketry uses flat ribbon-like elements to allow for a tight interlacing, and elaborate patterns can be formed in the tightly locked closed plane. The width of flat elements produces a stronger visual impact both of color patterns and of those which evolve from structural variations.

Plaited work is the same as braiding in that the structure is basically diagonal with one set of elements serving as both warp and weft in an over-and-under interlocking process. Terminology regarding plaiting varies considerably, and some forms have multiple sets of elements. All plaiting can be satisfactorily worked in metal. Such changes as plain weave (8-4) which is sometimes called checkerwork, basket weave, and twill weave (8-5) can become important design elements, although many complex and interesting patterns have been developed in basketry that are not used in braiding or weaving, especially those based on radial forms. The techniques of caning and rushing described later in the chapter are related to plaiting, although in caning the pattern is based on openwork with six sets of elements, and in rushing the traditional methods of forming the design involve a single continuous element.

Braiding, as we have seen earlier, can be accomplished with either round wire or flat strips and the same holds true for the process in the round used in basketry. Usually, the elements are set over one another for the bottom of the form, and as the sides are raised the braided interlocking takes place, sometimes in curves for rounded forms and sometimes turning in on themselves to form the corners of angled forms. In metalwork, the techniques offer unlimited possibilities for jewelry and containers as well as larger structures.

8-4. Plaiting is the basketry term for the same process of interlocking elements in an over-and-under sequence that is used in braiding. The structure shown here is plain weave in strips of sterling-silver sheet.

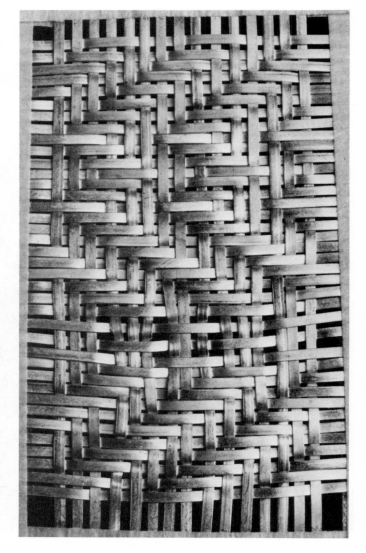

8-5. This example of plaiting is fine silver cloisonné wire in a twill weave.

TWINING

The technique of twining differs from any yet discussed, although it resembles weaving in its use of two dissimilar sets of elements, sometimes referred to as warp and weft. In basketry, the warp consists of a system of posts or spokes which may be assembled in a radial plan (see 8-7) or a parallel sequence, most often equally spaced, although this is not a structural necessity. When working in the round, uneven spacing results in a somewhat faceted surface, while even spacing with the elements quite close to each other will produce a smooth form with an even surface. Even drastically irregular spacing will not interfere with the basic integrity of the structure.

The weft, sometimes called the lacing, is a continuous, usually double element which may be manipulated in a variety of ways as it moves in and out of the warp structure. The term *wicker* is applied when a single weft is woven over and under the warps around and around the form (8-6). The total effect of this twined form of wickerwork is the same as a woven structure made with rigid materials which force the weft to move dramatically over or under each warp element. The term *twining* refers more specifically to the use of a pair of wefts which are crossed before twisting around each warp element (8-7) to produce a locked horizontal structure.

In basketry, the warp elements are only fastened at the base, which can be done by soldering in metal. The upper ends of the posts are left free for easy manipulation of the weft which is usually worked over the tops of the posts and then pushed down into position in the structure. This eliminates the need for raising and lowering the warp elements and allows the weft to travel freely without any need for constant pulling through the warp system. It is this feature of the process which makes it ideal for use in metal, since there is no trauma caused by the constant bending back and forth of the warp posts, and the weft can easily be kept free of kinks and bends if it is worked from a spool or bobbin.

The basic twining process lends itself readily to many variations, both in selection of materials and in weft pattern. The container in 8-6 (as well as the beads by the same artist in 8-2) are really one form of wickerwork in which a single weft is wrapped around each warp post. The rows of weft sit tightly against each other on the back of the warp posts, but spaces occur in the intervals and on the front of the warp because of the doubling of the weft as it twists behind the warp.

8-6. Silver container by Steven Brixner, constructed in the technique of twining. A single-weft element of 26-gauge fine-silver wire is twined continuously in a spiral pattern around evenly spaced vertical warp posts of 12-gauge round wire which have been first permanently fastened to the base. The base, rim, and upper terminal are fabricated of sterling-silver sheet. To secure the top edge, a sheet-metal ring is used, with holes drilled to receive the individual warp posts, which are soldered permanently after the twining is complete.

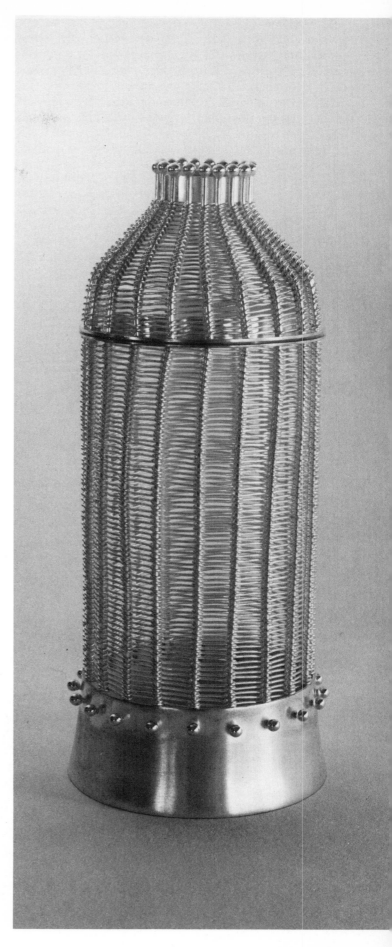

The normal twining system employs a double weft which is crossed (half turns) or twisted (full turns) before interlacing with the warp posts (8-7). Variations can include lacing around each warp post, around each two warp posts, a staggered progression of these lacing patterns as the twining proceeds upward, lacing with doubled pairs of wefts, with colored wefts, and so on.

The warp in metal can be shaped in curves impossible to achieve in the usual materials for twining, such as reeds or willow and bamboo splints, as it can be bent into place before or during the twining (8-7c). Interesting effects can be achieved with combinations of round, flat, or colored wires for the weft, and, if flat strips of flexible material are used, patterns resembling splintwork and wickerwork can be achieved.

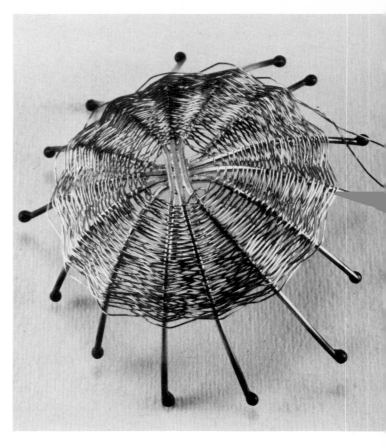

8-7a. Twining, step one. In this 18-karat gold piece, the warp elements, six strands of 16-gauge wire divided into two groups of three, have been soldered together in the center in an evenly spaced radial pattern. The twining is accomplished by a single length of 30-gauge wire doubled in half. In basketry with reed, a cross or x is usually made over the center section of the spokes, to hold the twining in place, but wire can be tacked in place with solder, or hooked securely around one of the spokes to begin.

8-7b. Twining, step two. Raising the basket is dependent on the way in which the posts are curved upward. The twining merely continues around the set form. Note how the posts have been finished by beading the metal with a torch.

Additional warp elements may be added to extend the form. The twining is then worked back and forth across the warp with the shape being controlled by a slight expansion of the warp and then a sharp compression, accompanied by an elimination of some warp ends at the terminal point. This twining process is used to construct small hollow forms in the brooch in 8-8, which are actually continuous in their structure. The warp elements move from one cylindrical area to the adjoining ones and are increased as necessary to provide sufficient warp for each new form. The weft is not continuous, but is worked generally from the base to the upper edge where the ends are forced back into the structure beside a warp element to form a permanent lock without the need for heat or solder. The twined section of the pendant is finished as one piece and is remarkably strong despite the delicacy of the material. The twining is done very tightly, which accounts for the durability of the form. It cannot easily be pushed out of shape or altered. The back of the pendant is a sheet of silver to which both chains and bezel-set stones are fastened. The basket form is placed on top of the plate and "sewn" into its permanent location with a supplementary length of wire. The total size of the twined part is a mere one-and-one-half inches by two inches, exclusive of the chain —an extremely delicate piece of jewelry.

8-7c. Twining, alternative form. Here posts in the same basic structure have been given an inward and then an outward curve to create a dramatic container form.

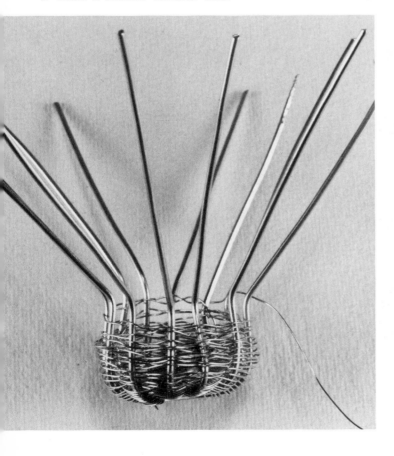

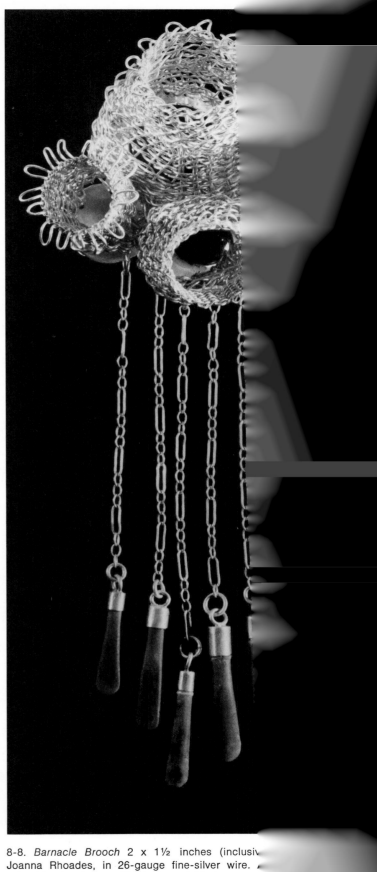

8-8. *Barnacle Brooch* 2 x 1½ inches (inclusiv̶
Joanna Rhoades, in 26-gauge fine-silver wire. ̶
twined basket forms with shells and Mexican opa̶
into a single piece by the addition of warp for n̶
is twined with noncontinuous weft. The piece ̶
with wire to a base plate of sterling-silver sheet

COILING

The third major basketry process is coiling, which is primarily a single-element stitching process worked continuously over a wrapped core. The core provides the bulk and the dimension of the form, and the wrapping element controls the structure and shape of the piece as work proceeds.

In fiber, the core is generally some kind of heavy material, such as reed, cording, rope, a group of multiple yarns or even bunches of grass, pine needles, or other natural materials of a pliable character. Because of this pliability, contemporary artists are using the process to make highly convoluted surfaces on all types of forms—flat, relief, symmetrical, polymorphous—and in every scale from one-inch containers to large wall reliefs. These complexities are possible because the structure is not hindered by a predetermined warp or a particular frame or implement. The form is able to evolve as the structure grows. The core can be manipulated into spirals, straight lines, even vertical loopings, yet still continue to build a self-sustaining structure.

8-9. Coiled basket in sterling-silver wire with seashells by Anita Fechter. The container is 10 x 10 x 11 inches and utilizes the coiling technique in a free-flowing and highly decorative manner. The core is 12-gauge wire with a weft of 18-gauge wire. (Photograph courtesy of the American Crafts Council.)

All of these possibilities exist in metal, including the great variation of scale. The core in metal coiling is generally wire or a group of wires, although it could as well be thick-walled plastic tubing of the type found in electrical and electronic cables. A very tiny container might have a flexible core of heavy cord wrapped with thin-gauge wire (8-11). A larger structure might have groups of twisted wire elements as a core (see 8-10a) or groups of already coiled elements used as a core (see 8-3). Although the wrapping is usually done in such a way as to totally cover and obscure the core material, the structure can be achieved with a less dense wrapping that allows a decorative core to show to advantage.

In metal, the process of coiling can be used in a way not possible in fiber or plant material. The stiffness of wire and the inherently strong structure of the coil that is formed makes it possible to remove the core and use the empty coils as elements. The wire will remain in its dense coiled form and continue to be self-sustaining even after the core is removed. Either on or off the core, metal coils can be stretched out like springs providing the craftsman with a wide range of decorative effects. Although this technique uses a core and coiling element, it is really closer to the technique of wrapping than to the basketry technique of coiling, which involves the stitched *connection* between two rounds of core material.

In coiling, the weft element must be extremely pliable. Grasses, rushes, or raffia have been traditionally used, and many forms of yarn have also been employed. In metal, very thin wire behaves in the same manner as yarn, and can even be used in a needle as is often done in basketry. The only problem is the knotting and kinking of the wire, which can be avoided by using short lengths and taking care to keep it untangled as work proceeds. As new lengths of wire are needed they can be hidden under the last coils of the old length, making joining a very easy procedure. Since grasses come in short lengths, primitive basket makers had to resort to the same tactics. The need to use short weft ends to simplify the stitching process can lead to the development of complex patterns through frequent changes of color or alternations of dark and light elements.

The basic procedure for making either a flat form or a dimensional one is the same. The core material is first wrapped a short way and then bent in on itself to form the beginning of the shape. Although they are most often circular (8-10a), the bases can also be created in elongated oval shapes by folding a greater length of core in on itself. Next the weft is continuously wrapped around the core (without a needle if working in thick wire) in a stitch and pattern determined beforehand. The simplest form of coiling is shown in the photographs—it is merely wrapping that is stitched through two rows after a specific number of wraps, in this case four. The end of the wire is inserted directly under the preceding coil and pulled through with the fingers. The rows are pulled tightly together to produce a stable form.

8-10a. Coiling, step one. To begin, the core of multiple strands of copper wire is wrapped a few times and then bent back on itself. The free end of the wrapping was tucked in to secure it. The wrapping stitch used here calls for wrapping the 28-gauge coated copper wire around two rows after every four wraps on a single row. There are many other stitches that may be used to hold the coils together. The free end of the wire is shown here at the left, as one of the securing stitches is being made.

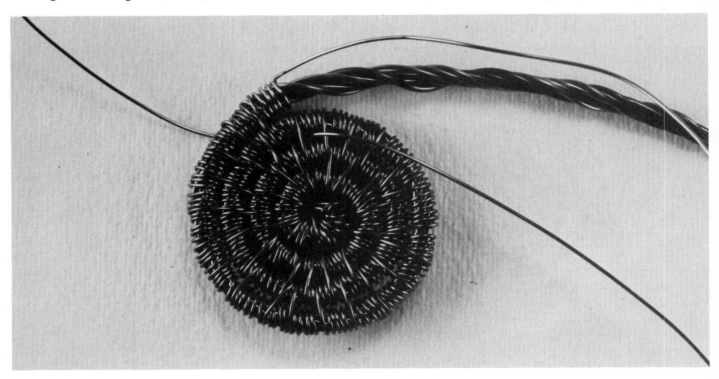

The materials are pliable enough for the basket maker to exercise complete control over the form with a minimum of physical exertion. The raising is achieved by the way in which the coils are spiraled up on one another (8-10b), and extremely complex forms can be achieved in a freehand manner that takes advantage of occasional openings in the structure to add to the decorative quality (8-9). Finishing a coiled form in fiber usually involves the tapering of the last piece of the spiral to achieve a level top margin. This can be accomplished in metal by filing the core material to a long flat taper before wrapping.

8-10b. Coiling, step two. The basket form is raised by slowly spiraling the core material upwards in the desired curve, and securing it as coiling progresses.

8-10c. Coiling, step three. When the basket has reached the desired height, the core material is tapered to finish off evenly. In metal this effect is achieved by filing the core into a long flat taper. Note the use of the awl (a steel rod with a rounded point set into a wooden handle) to help in separating the coils temporarily for the insertion of wire in the joining stitch.

8-11. Top view of an open coiled basket. This one-inch diameter basket is intended to house a finger ring. The core is a sturdy fiber cord, and the wrapping is done with 36-gauge red-coated copper wire.

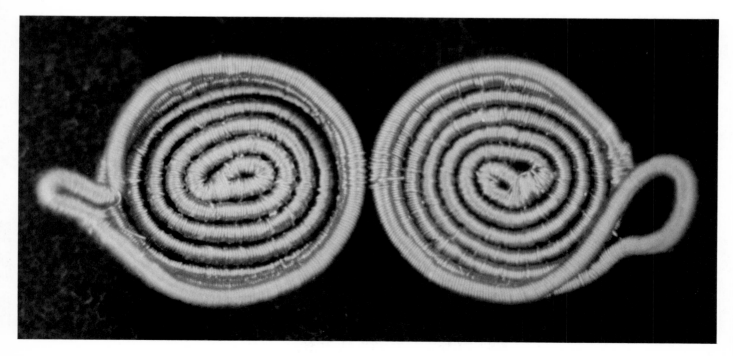

CANING AND RUSHING

Associated with basketry because of the materials and their woven effect are two interlacing techniques usually connected with the making of furniture, especially the seats and backs of chairs. Both of them are adaptable to work in metal, although the rigid frames required to support the structure when cane and rush are used are best abandoned in favor of a more flexible device. Flat metal strips achieve a well-defined pattern and a closely interlocking structure that most closely resembles the traditional product. Round wire produces more open effects that are structurally stronger than the same construction would be in fiber.

Caning is accomplished through a series of multiple elements which are first layered *without interlacing* in a horizontal/vertical grid pattern, and are then interlaced by two separate sets of diagonal elements that lock the entire structure. The standard caning pattern involves six separate sets of elements, each evenly spaced, which yield an overall effect of octagonal openings within an omni-directional surface. These elements are arranged in a very particular sequence which guarantees the structure and the pattern (8-13). Some variations are possible without damage to the integrity of the structure, but the pattern is immediately altered in effect by even the slightest deviations.

8-12. The basic pattern in caning involves six separate sets of elements, distinguishing it from the single set of elements in plaiting. The horizontal and vertical grid is laid down first before the diagonal elements are inserted, so that the dimensions of the form are set, and it cannot grow continuously along the warp direction. This sample is done in round wire of 26-gauge fine silver. All caning and rushing samples are by Brigid O'Hanrahan.

8-13a. Caning, layer one. Vertical elements are arranged in parallel fashion, evenly spaced. The sample in this series is constructed of fine-silver cloisonné wire.

8-13b. Caning, layer two. Horizontal elements are arranged in a parallel system with spacing equal to the vertical elements. Note that the horizontal layer is merely laid on top of the vertical layer, not interlaced.

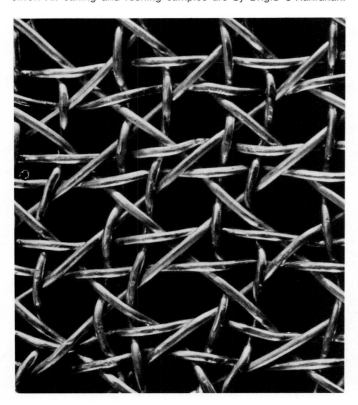

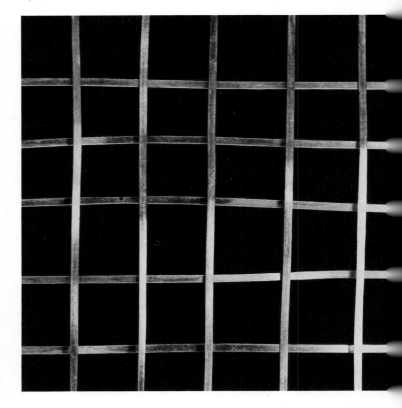

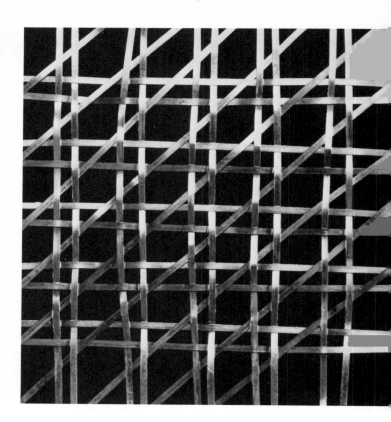

8-13c. Caning, layer three. Another series of vertical elements is set directly over the two previous layers, with elements arranged slightly to one side of those in layer one. Again, the elements are not interlaced.

8-13d. Caning, layer four. This horizontal layer is the first one to be interlaced, The elements are set as close as possible to the elements in layer two, and are interlaced with the vertical elements of layers one and three, as shown.

8-13e. Caning, layer five. Starting at the upper right corner, a series of elements is woven diagonally across the entire structure in a sequence of under two vertical elements and over two horizontal elements. This is the first layer in which all of the elements are not the same length.

8-13f. Caning, layer six. Starting in the upper left corner, another series of elements is woven diagonally across the structure in the opposite sequence of over two vertical elements and under two horizontal elements.

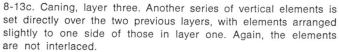

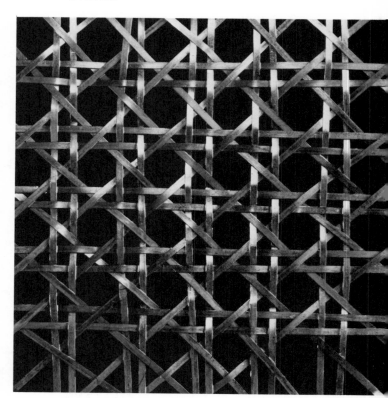

The very simple deviation of reversing the over and under sequence of the diagonal elements in relation to the interlaced structure of the vertical and horizontal elements creates a distinct warping of the pattern (8-14). The entire surface is slightly more textural, and the diagonals take on a curvilinear pattern which has a more sinuous quality than the usual straight lines and diminishes the importance of the openings in favor of the diagonals.

An alteration in the type of materials used also produces noticeable visual changes in the pattern. The use of round wire in place of flat ribbons of metal results in a more textural surface and a lessening of the significance of the pattern formation (see 8-12). Done on a large scale with large-diameter wire the texture of this particular variation would be quite powerful—far more so than its pattern. In a combination of flat strips and round wire the pattern dominates but the change in materials produces a slight change of surface quality and light reflection.

The rigid frames of chair seats provide the supporting structure during caning and remain as the frame in the finished product. The frame requires that the caning material be soaked in water to make it sufficiently flexible and pliable to survive the interlacing procedure. Such a convenience is not possible in metal, although thin and malleable metal should certainly be selected for the best working properties. To use the metal in the same way as cane is difficult because it is uncomfortable as well as damaging to the surface of the metal to pull the interlaced strips along through the structure when all the layered elements are fastened tightly at both ends. A much more convenient method is to fasten the ends of all the elements between strips of masking tape. This holds them securely in relation to each other while providing a flexible frame to make the interlacing easier. The finished piece can be fastened to a rigid frame if desired by riveting, nailing, soldering, lacing, or any other suitable method.

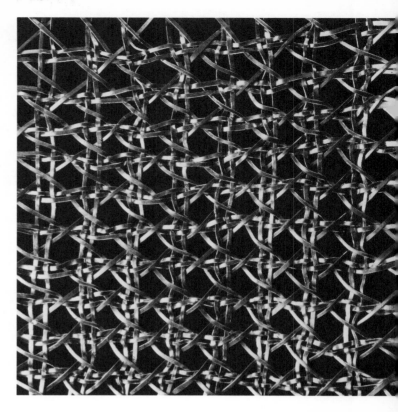

8-14. A caning variation. The simple pattern deviation of reversing the over and under sequence of the diagonal elements in relation to the interlaced structure of the vertical and horizontal elements creates a more textured surface, with curved rather than straight diagonals, which diminishes the visual effect of the openings. This sample is done in fine-silver cloisonné wire.

8-15. A combination of flat strips and round wires of 26-gauge fine silver used to create a caning variation.

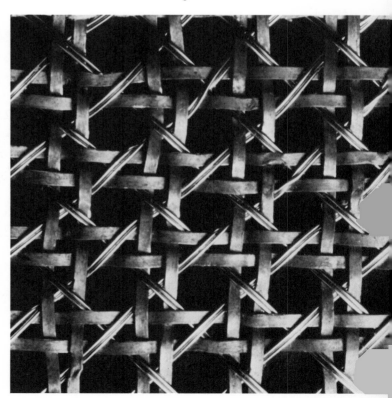

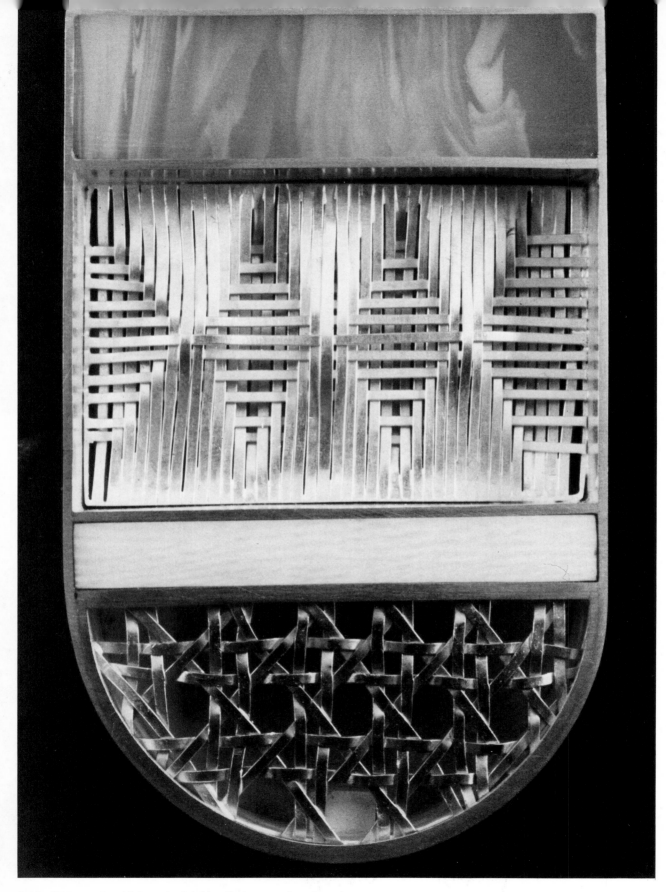

8-16. Silver and gold brooch, inlaid with ivory and amber, by Brigid O'Hanrahan. The center area uses a dimensional rushing pattern in fine-silver wire, which can be viewed from both sides. The lower container area uses a facing pattern of caning done in flat-gold wire.

Rushing is used almost exclusively for the making of chair seats in natural rush, raffia, twine, or rope. It is an interlacing process which is constructed over a square or rectangular frame and is based on a figure-eight motion. In contrast to caning, which uses many sets of elements, rushing is a single-element structure, with the material being spliced or knotted together as additional length is required. The work is begun in all four corners simultaneously and progresses toward the center, wrapping around the frame after every interlacing. Done in this manner, it is not very well suited for work in metal because of the constant manipulation of a single length, which causes work hardening and unwanted kinks, bends, and twists in the metal. It is possible, however, to build up the rushing patterns using separate elements in the form of flat strips. The thickness of the metal produces a more pronounced dimensional quality to the pattern than develops when soft, fibrous materials are used. Because the interlacing is not a simple balanced structure of over one, under one the pattern is totally different on the reverse side, as illustrated in 8-17a and b.

The basic procedures of a rush interlacing done with two sets of elements consist of the creation of a parallel warp of an uneven number of ends and the weaving in of weft elements in a symmetrical pattern of increasing and decreasing floats, as shown.

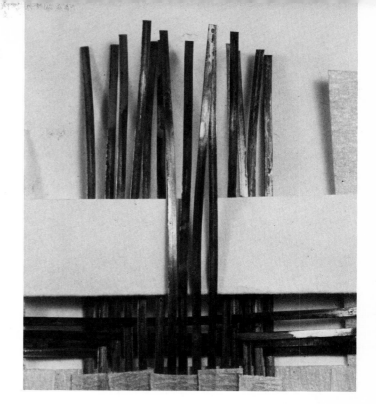

8-17b. Rushing, step two. The weaving process is facilitated by separating the vertical strips for each insertion with a strip of stiff paper.

8-17c. Rushing, step three. Separate horizontal weft elements are woven across the warp structure in a constantly increasing sequence until the center is reached, at which point the sequence is reversed. The first weft element is woven over the first warp, under all other warps, and over the last warp. Each successive strip in the progression increases (by one on each side) the number of outer vertical strips it crosses over, and decreases the number of central strips it crosses under. This is continued until the weft element is woven over all warps except the center one. The sequence is then worked in reverse to produce a symmetrical pattern. The cross strips are placed close together and held in position at the sides with strips of masking tape.

8-17a. Rushing, step one. A warp of an uneven number of vertical strips of 30-gauge brass is arranged in a parallel sequence under a strip of masking tape. The elements are spaced at intervals equal to the width of the warp strips. Rushing is traditionally done with a single unbroken length of material interlaced around a frame in a figure-eight pattern; with metal two separate sets of elements are used because a single continuous wire would become kinked and knotted.

Many other sequences are possible, some of which are less dramatic in sculptural quality and which place more emphasis on surface pattern. All kinds of twill sequences can be used, which involve taking the weft over and under multiple warps. If each weft element shifts its sequence by one warp, diagonal and herringbone patterns of great complexity can be produced. The flat strips of metal can be pushed very close together, producing a highly textured surface that would be effective on both small and large-scale.

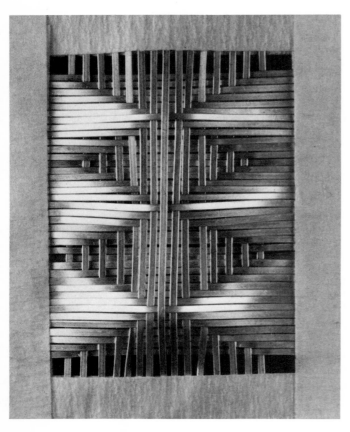

8-18a. A rushing pattern is made with lengths of fine silver cloisonné wire. In rushing the entire pattern can be repeated any number of times for the total length of the warp.

8-18b. The reverse side of the rushing pattern shows clearly the dimensional quality of the technique produced by using metal. Many variations of square, rectangular, and diamond patterns can be created.

9/Contemporary Work, A Pictorial Survey

In collecting the work of contemporary artists it has been exciting to discover why and how individuals from many parts of the world have come to use textile techniques directly in metal. The *why* is usually the result of necessity; normal metalworking processes simply were not adequate to produce the desired form, flexibility, pattern, texture, or weight. The *how* varies enormously, with some individuals actually re-inventing a technique in metal simply because they did not have prior knowledge of the desired textile processes. Others approach the problem by incorporating only as much as is necessary of an already familiar textile technology to meet the demands of a different material. Even these adaptations have an individual character, with different people producing the same result by varying means.

It is interesting to note that almost all of the artists whose work is illustrated are jewelers or sculptors; only one is a weaver. This may be the result of a need felt by jewelers to expand their resources in the direction of flexible and lightweight structures. That the use of metal has not been more fully explored by fiber artists is best explained, perhaps, by the abundance of other materials at their disposal and by the constantly expanding choice of man-made fibers with structural potential. The wealth of inventive solutions shown in this book suggests that there are indeed textiles that could well be made directly in metal and that textile artists may soon begin to explore these possibilities.

These photographs illustrate finished works in a wide variety of textile processes, many of which have been made with methods devised by the artist to accommodate the particular characteristics of metal. The inventiveness of the technical solutions is extraordinary, and I have included the artist's own description of the process wherever possible.

9-2. Woven necklace by Valerie Clark. Constructed without the support of board and pins, the rigid contoured outer form is made of wire soldered at the corners; the alternating wire and sheet-metal warp elements are soldered to the frame at the upper edge but left free at the lower edge to facilitate weaving and the incorporation of wooden beads. After the weft has been woven horizontally, it is soldered to the frame at both ends. The free warp ends are fastened mechanically at the bottom edge and ornamented with the addition of feathers.

9-3. A pendant by Linda C. Bell, 3½ x 3 inches. The wing extensions are round silver wire in a woven construction, while the center area uses a strip of card weaving in silver and brass as an ornamental device supported by a three-dimensional structure of sheet silver and tubing.

9-1. Wall hanging, 18 x 45 inches, loom-woven copper warp and weft in various gauges and colors, by Mel Someroski.

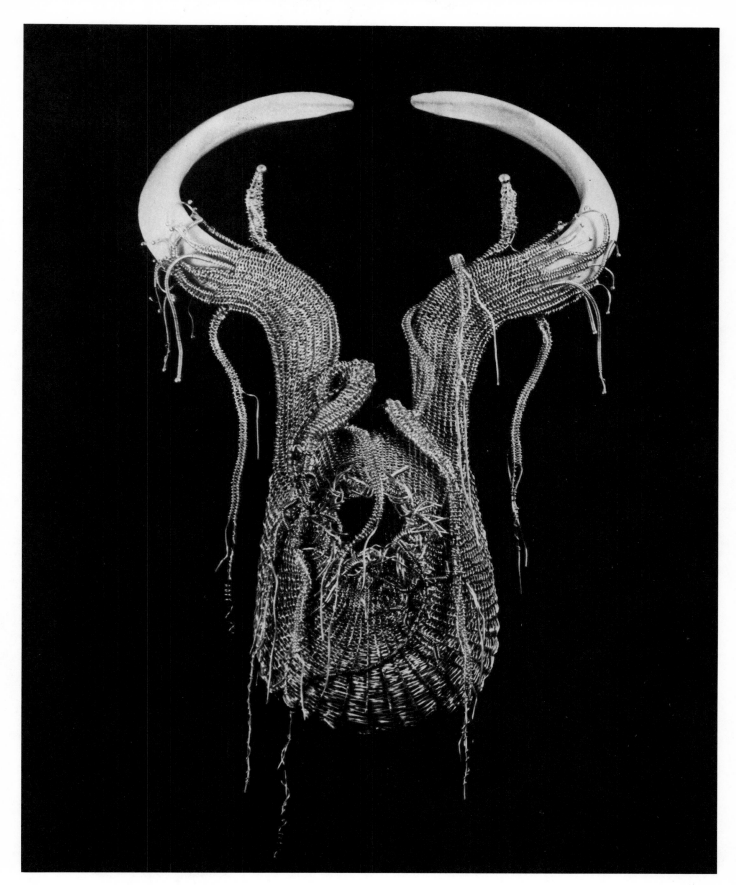

9-4. Woven neckpiece, fine-silver and gold-filled brass wire, boar's tusks, by Mary Lee Hu. The pouch is woven over a rigid warp in the same manner as a basket would be made. The unused warp ends are wrapped and beaded to give an elegant finish.

9-5. Necklace of flat-knitted mesh in silver, by Richard Anderson. He describes the technique he used to make all his pieces shown here as follows: a flattened wire spring is made which can be interlocked with similar elements to form a mesh which is flexible in its length while remaining rigid in its width. The cylindrical pieces shown are made by producing a long length of loops and locking them together in a continuous spiral pattern. Color changes are possible by varying the metal with which the loops are made.

9-6. Tubular-knitted bracelet by Richard Anderson. Eighteen loops are used in the circumference of a tube which is flattened to produce the double fabric. The ends are joined together with silver rivets.

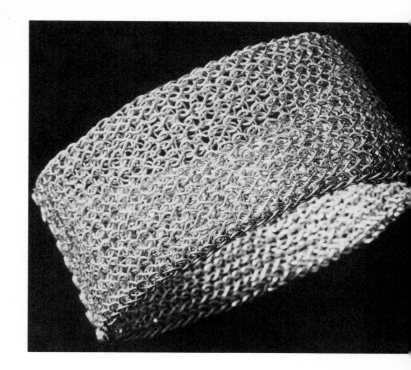

9-7. Tubular-knitted necklace by Richard Anderson. Alternating bands of silver and copper produce a pronounced light and dark pattern.

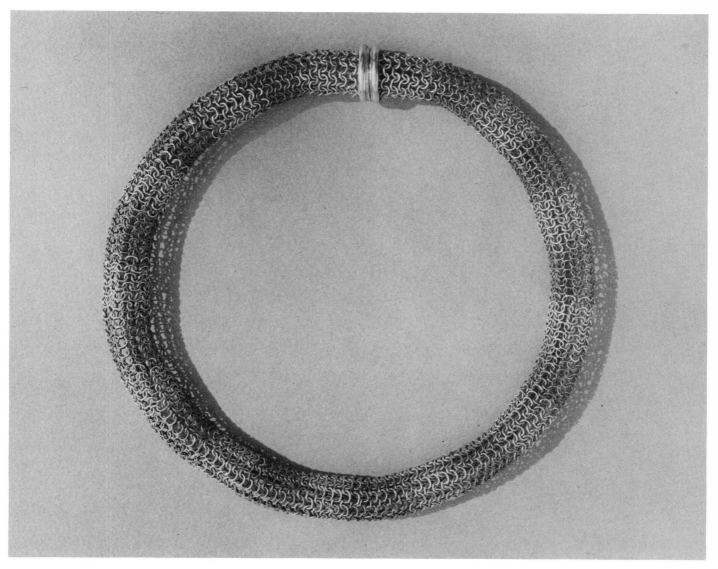

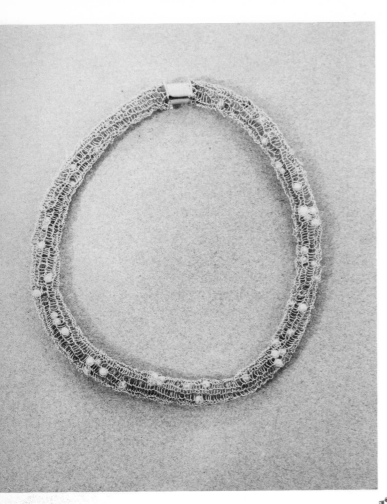

9-8. Spool-knitted necklace in silver gilt wire with white and black pearls by Charlotte de Syllas. Collection of The Worshipful Company of Goldsmiths, London. Photograph courtesy of Hornsey College of Art.

9-9. Spool-knitted collar, 18-karat gold, by Jacqueline Mina. The tubular form is knitted on a spool with four widely spaced pins and then twisted to give a spiral effect.

9-10. Necklace in gold sheet and wire and blue linen, which uses a knotted structure of pre-Columbian origin. The upper edge is a spool-knitted tube of 18-karat wire stuffed with a second tube of linen thread. The 14-karat sheet circlets increase in size to complement the curvature around the shoulders and are put together with a series of half-hitch knots in double and single strands of linen, although the knotting could equally well be done in strong, malleable metal.

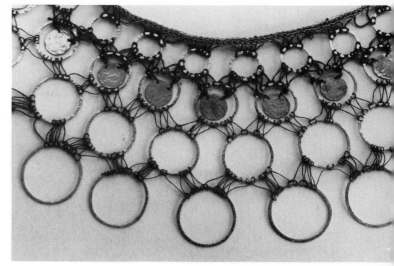

9-11. A Moebius-type wire structure in brass by Ruth Asawa, 24 x 33 inches in diameter. Photograph by Laurence Cuneo.

9-12. *Neckpiece #13,* fine silver, fine gold, gold-filled brass, natural ruby, and pearls, by Mary Lee Hu. The outer band and neckpiece is a single tube of silver wire crocheted in a dense structure. The central image of a peacock is composed of linked elements in a variety of dimensions, patterns, and colors. The joinings are made with textile processes, so that the area retains its flexibility.

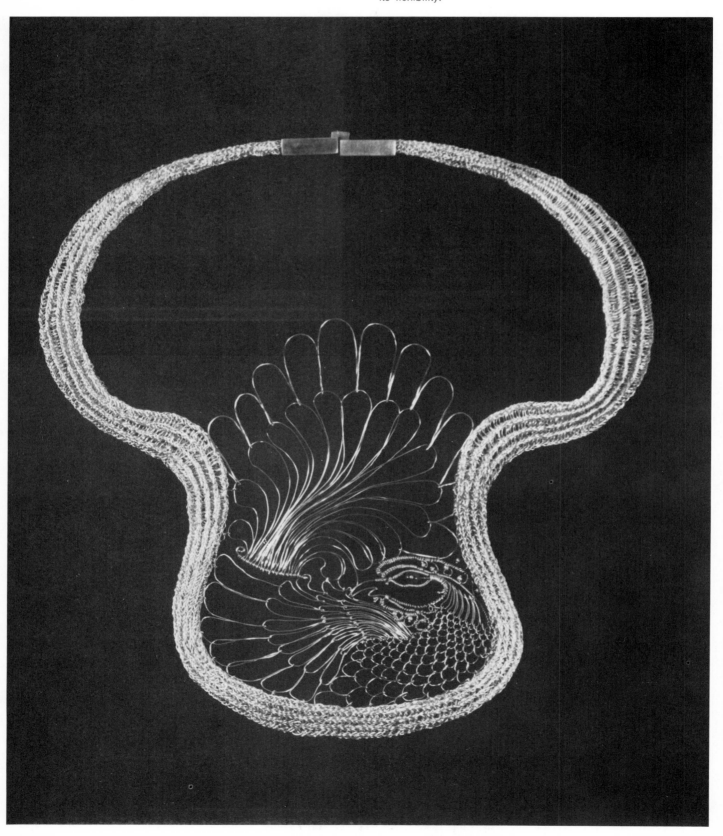

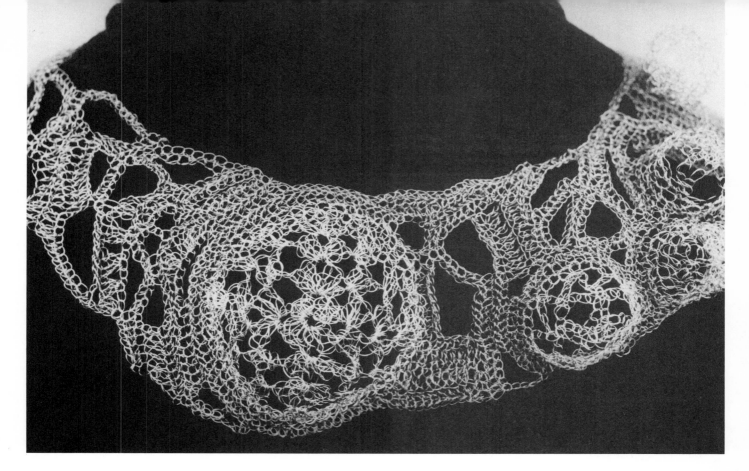

9-13. Front view of a large crocheted neckpiece by a student. Multiple colors and sizes of wire are used to create an intricate and highly patterned surface. The circular areas on the right are actually three-dimensional forms, while others stand away from the body at varying levels. The form continues around to the back where a large circular pattern is the focal point. The necklace is buttoned at one shoulder with a simple chain-stitch loop which slips over a small spherical shape.

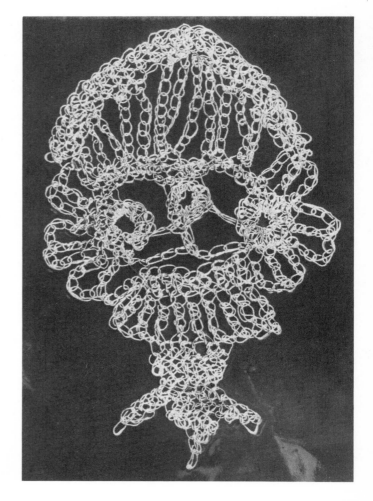

9-14. Wall panel by Janis Rosenthal. After being crocheted into a tree form, brass wire is fused to the surface of an enamel panel, a process which does not destroy either the structural detail or the color of the metalwork.

9-15. Copper neckpiece, 4 x 8 x 7 inches, Anita Fechter. Fashioned completely from braided industrial tubing, the undulating form takes great advantage of the natural expandability and compressibility of the material. Sections and edges are laced together, and the whole piece is electroplated to stabilize the form and add small textural nodules along the edges.

9-16. Detail of a string of beads by Steven Brixner. Hand-carved tubes of ivory alternate with lengths of square braid in fine silver and discs of silver and shell.

9-17. Ring, 14-karat yellow gold wire and pearls, by Wann-Hong Liu. A simple three-strand braid of coiled wire forms the ring part, while the setting carries out the design in further coils of the same coiled wire.

9-18. A simple form of coiling is shown in this four-knot bracelet by Bob Lee, available in copper, silver, and 14-karat gold, and based on the elephant-hair bracelets worn by African hunters for good luck. Courtesy of Hunting World, New York (U.S. Patent #224257).

9-19. A party bracelet by Wann-Hong Liu in sterling-silver wire uses the simple process of coiling and rewinding with coiled forms.

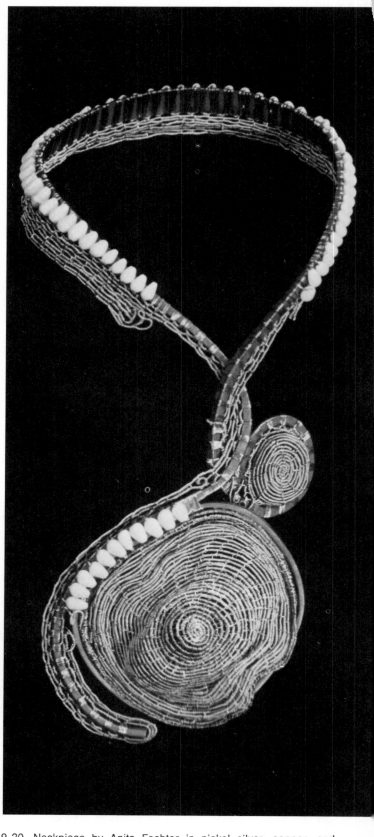

9-20. Neckpiece by Anita Fechter in nickel silver, copper, and shells. The circular areas are formed in coiling technique with an embellishment of heavy forged wire that forms the structural basis of the necklace. The thin nickel silver wire is coiled around a core of 16-gauge copper wire, and the rows are connected by coiling through holes made in the previous row with a sharp instrument like a compass point or an etching needle.

9-21. Twined pendant by Joanna Rhoades. Coated 28-gauge copper wire is twined over a free-form radial warp of the same material forming a shallow bowl form. The ends are finished by decorative coiling. Additional warp elements are added on one edge of the bowl to extend the form into a semi-figurative image, with the shape being controlled by the spreading or narrowing of the warp, and some of the warp ends are eliminated at the terminal point.

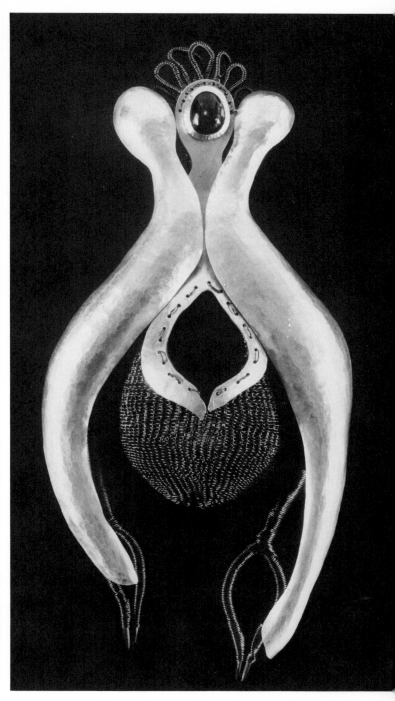

9-22. Pendant in hammered sheet silver by Thelma Coles. The hollow basket is formed by twining 28-gauge red coated copper wire and stitching it onto the frame. Decorative elements are created by wrapping with red wire.

Appendix:

Suppliers

TOOLS

Allcraft Tool and Supply
22 West 48th Street, New York, N.Y. 10036 (N.Y. salesroom)
204 North Harbor Blvd., Fullerton, Cal. 92632
(Cal. salesroom)
215 Park Avenue, Hicksville, N.Y. 11801 (Mail orders)

Anchor Tool and Supply Company
231 Main Street, Chatham, N.J. 07928

California Crafts Supply
1096 N. Main Street, Orange, Cal. 92667

William Dixon Company
Carlstadt, N.J. 07072

Friedheim Tool Supply Company
412 West 6th Street, Los Angeles, Cal. 90014

Paul Gesswein and Company, Inc.
235 Park Avenue South, New York, N.Y. 10003

I. Shor Company, Inc.
71 Fifth Avenue, New York, N.Y. 10003

Swest, Inc. (Southwest Smelting and Refining Company)
10803 Composite Drive, Dallas, Texas 75220
118 Broadway, San Antonio, Texas 78295
1725 Victory Blvd., Glendale, Cal. 91201

PRECIOUS METALS

T. B. Hagstoz and Son
709 Sansom Street, Philadelphia, Pa. 19106

Handy and Harman
850 Third Avenue, New York, N.Y. 10022
4140 Gibson Road, El Monte, Cal. 91731
1900 Estes Avenue, Elk Grove Village, Ill. 60007
Frank Mossberg Drive, Attleboro, Mass. 02703

4402 West 215th Street, Cleveland, Ohio 44126
17000 West 8 Mile Road, Southfield, Mich. 48075
1234 Exchange Bank Bldg., Dallas, Texas 75235

Martin Hannum, Inc.
810 South Mateo Street, Los Angeles, Cal. 90021

Hauser and Miller Company
4011 Forest Park Blvd., St. Louis, Mo. 63108

C. R. Hill Company   (silver and gold cloisonné wire)
2734 West 11 Mile Road, Berkeley, Mich. 48072

Hoover and Strong   (gold only)
119 Tupper Street, Buffalo, N.Y. 14201

I. Miller, Inc.   (silver only)
304 Colonial Arcade, Cleveland, Ohio 44115

Swest: see address above

Wildberg Brothers Smelting and Refining Company
(gold only)
635 S. Hill Street, Los Angeles, Cal. 90014
742 Market Street, San Francisco, Cal. 94102

COPPER AND BRASS

Belden Corporation   (copper wire)
Chicago, Illinois

T. E. Conklin Brass and Copper Company, Inc.
324 West 23rd Street, New York, N.Y. 10011

Revere Copper and Brass, Inc.
230 Park Avenue, New York, N.Y. 10010

PEWTER

White Metal Rolling and Stamping Corp.
80 Moultrie Street, Brooklyn, N.Y.

Recommended Tools and Shop Equipment

(Starred items are the most important)

BASIC HAND TOOLS

Cutting Tools:
*Large scissors (paper cutting)
 Plate shears with straight and/or curved blade
*Wire cutters (diagonal or end cutting)

Bending Tools:
*Jewelers pliers (chain and round nose)

Forming Tools:
*Rawhide mallet
*Ball pein hammer
 Planishing hammer
 Rivet hammer
 Sandbag
 Lead Block
*Steel block or anvil

Miscellaneous:
*Files (assorted)
*Hand drill and twist drills #50–65
 Wooden hand clamp
 Pointed jeweler's tweezers
 Drawplate and drawtongs

FINISHING AND POLISHING EQUIPMENT

*Emery paper #300–600
 Fine steel wool
 Felt buffing stick
 White diamond compound
 Brass wire brush
 Hollow scraper
 Burnisher

SHOP EQUIPMENT

Torch (Prest-O-Lite, Bernzomatic) with
interchangeable tips.
Turntable with asbestos pad
Flame-proof pot for pickle
Small table vise
Hand rolling mill
Steel stakes and anvils

Comparative Sizing of Hooks and Needles

COMPARISON OF SIZES OF CROCHET HOOKS AND NUMBERING SYSTEMS

mm.	steel (U.S.)	steel (U.K.)	aluminum or plastic (U.S.)	aluminum (U.K.)	wood (U.S.)	Giant plastic (U.S.)	New International numbering
.60		6½					.6
	14	etc.					
.75	13	5					.75
	12	4½					
1.00	11	4					1
	etc.	3½					
1.25	7	3					1.25
1.50	6	2½					1.50
1.75		2					1.75
		1½					
2.00	5	1		14			2
	4	1/0					
2.50	2	2/0		12			2.5
	etc.						
3.00	00	3/0	D or 3	11			3
3.50			E or 4	9			3.5
4.00			F or 5	8			4
4.50			G or 6	7			4.5
5.00			7	6			5
5.50			H or 8	5			5.5
6.00			I or 9	4	10 or I		6
7.00			J or 10	2			7
			K or 10½				
8.00					11 or L		
9.00							
10.00					13 or M		
11.00					15 or N		
12.00							
13.00					16		
14.00							
15.00							
16.00						Q	
17.00							
18.00							
19.00						S	

COMPARATIVE SIZES OF U.S. AND BRITISH KNITTING NEEDLES

U.S.	BRITISH
00	14
0	13
1	12
2	11
3	10
4	9
5	8
6	7
7	6
8	5
9	4
10	3
10½	2
11	1

A COMPARISON AMONG HOOKS OF VARIOUS MATERIALS AND SIZES

Steel	14-4 3 2 1 0 00
Aluminum	B C D E F G H I J K
Plastic	D E F G H I J K
Plastic	3 4 5 6 7 8 9 10 10½
*Afghan	B C D E F G H I J
Bone (tapered)	1 2 3 4 5 6
Wood	I L M N
Wood	10 11 13 15 16
Giant Plastic	Q S
diameter	¼" ⁵⁄₁₆" ⅜" ⁷⁄₁₆" ½" ⅝" ¾"

*Afghan hooks—numbers correspond to knitting needles of same letter.

Abbreviations

KNITTING

- k = knit
- p = purl
- yo or o = yarn over
- sl = slip stitch
- psso = pass slip stitch over
- k2tog = knit 2 stitches together
- p2tog = purl 2 stitches together

CROCHET

- c = chain
- ch st = chain stitch
- sc = single crochet
- dc = double crochet
- st = stitch
- sts = stitches
- tcr = triple crochet
- r = row

BOBBIN LACE

- T = twist
- C = cross
- TC = half stitch
- TCT = linen stitch
- TCTC = whole, double, or lace stitch

Melting Points

Metal	°C	°F
Aluminum	660.1° C	1220.29° F
Brass (yellow)	904.0	1660.0
Copper	1083.0	1981.4
Gold (24k)	1063.0	1945.4
(18k)	927.0	1700.0
Iron	1535.0	2795.0
Lead	327.35	621.3
Nickel	1455.0	2651.0
Pewter	245.1	475.0
Silver (fine)	960.5	1760.9
(sterling)	893.0	1640.0
Tin	231.9	449.4

Comparative Systems of Metal Measurements

Brown & Sharpe Gauge	Thickness in Inches	Thickness In Millimeters	Ounce Weight of Copper Sheet per sq. foot
14	.06408	1.628	48
16	.05082	1.291	36
18	.04030	1.024	30
20	.03196	0.812	24
22	.02534	0.644	18
24	.02010	0.511	15
26	.01594	0.405	12
28	.01264	0.321	9
30	.01003	0.255	7.5
32	.00795	0.202	6

Reference Chart of Materials Suitable for Different Textile Techniques

	continuous single-element techniques (crochet, knitting)	noncontinuous single-element techniques (braiding, inter-linking, sprang, bobbin lace)	multiple-element techniques (weaving, basketry)
gold	24, 22 or 18 karat in a thin gauge is excellent, and does not have to be annealed. 14 karat in a soft alloy and thin gauge may be used but will require frequent annealing.	18 karat is sufficiently malleable; 14 karat in a soft alloy may be used; 24 karat may be used but will be too soft for many finished items.	Fine gold is too soft to be self-supporting unless interlaced with harder metals. Weaving can be done with as low as 14 karat if a sufficiently thin gauge is used (30).
silver	Use fine silver wire because it does not work harden rapidly and so requires a minimum of annealing.	Use fine silver wire; inter-linking and sprang can be done with sterling.	Sterling is good and will hold structure well but it is difficult to interlace in compact patterns. Fine silver sheet and wire are excellent.
copper	Wire is excellent in all gauges. Varnish on wires will not crack or peel if they are not heated, treated chemically, hammered, or rolled to change form.	See remarks for continuous techniques.	Wire in thin gauges may be used in loom weaving and is sufficiently flexible even for cardweaving and pinweaving. Sheet metal cut into strips of any width and gauge functions well.
brass	Wire may be too inflexible in medium gauges or break too easily in thin gauges; must be annealed.	Wire is good, especially yellow brass, except where too much stress is involved, as in sprang.	Good in strips of thin gauge cut for weaving and caning, but must be annealed. Holds a self-supporting structure very well.
pewter	Excellent if large gauge can be used, or if it can be drawn down to finer gauges for small-scale work. Never has to be annealed.	Wire in large gauge is good but too soft to hold structure well.	Cut strips are excellent in combination with harder metals.
other metals	Stainless steel (nautical) wire; iron binding wire	Iron binding wire can be used effectively since it is very flexible and does not work harden too quickly; stainless steel wire also works.	Large-scale architectural pieces can be done in aluminum, steel, bronze.

Bibliography

BASIC METAL TECHNIQUES

Bovin, Murray, *Jewelry Making,* Forest Hills, New York: Murray Bovin, 1967.

Bovin, Murray, *Silversmithing and Art Metal,* Forest Hills, New York: Murray Bovin, 1963.

Morton, Philip, *Contemporary Jewelry,* New York: Holt, Rinehart and Winston, 1969.

Untracht, Oppi, *Metal Techniques for Craftsmen,* Garden City, New York: Doubleday and Company, 1968.

Von Neumann, Robert, *The Design and Creation of Jewelry,* Philadelphia: Chilton, revised edition, 1974.

TEXTILES

American Fabrics Magazine, *Encyclopedia of Textiles,* Englewood Cliffs, New Jersey: Prentice-Hall, 1960.

Birrell, Verla, *The Textile Arts,* New York: Harper and Row, 1959; Schocken, 1973.

Emery, Irene, *The Primary Structures of Fabrics,* Washington, D.C.: The Textile Museum, 1966.

Held, Shirley E. *Weaving, A Handbook for Fiber Craftsmen,* New York: Holt, Rinehart and Winston, 1973.

WEAVING

Atwater, Mary, *Byways in Handweaving,* New York: Macmillan, 1954.

Black, Mary, *New Key to Weaving,* Milwaukee, Wisconsin: Bruce Publishing, 1957.

Blumenau, Lili, *The Art and Craft of Hand Weaving,* New York: Crown, 1955.

Crockett, Candace, *Card Weaving,* New York: Watson-Guptill, 1973.

Overman, Ruth and Smith, Lulu, *Contemporary Handweaving,* Ames, Iowa: The Iowa State College Press, 1955.

Regensteiner, Else, *The Art of Weaving,* New York: Van Nostrand Reinhold, 1970.

Specht, Sally and Rawlings, Sandra, *Creating With Card Weaving,* New York: Crown Publishers, 1973.

Weaving on a Card Loom, issued by the Educational Bureau of Coats and Clark, Inc., 430 Park Avenue, New York, New York 10022, 1955.

Znamierowski, Nell, *Step-By-Step Weaving,* New York: Golden Press, 1967.

KNITTING

Abbey, Barbara, *The Complete Book of Knitting,* New York: Viking, A Studio Book, 1971.

Book of Patterns and Instructions for American Needlework, Pamphlet #43, "Knitting," New York: Fawcett Publications, 1961.

Phillips, Mary Walker, *Creative Knitting, a New Art Form,* New York: Van Nostrand Reinhold, 1971.

Phillips, Mary Walker, *Step-By-Step Knitting,* New York: Golden Press, 1967.

Taylor, Gertrude, *America's Knitting Book,* New York: Charles Scribner's Sons, 1968.

Thomas, Mary, *Mary Thomas's Book of Knitting Patterns,* London: Hodder and Stoughton, Ltd., 1943.

Thomas, Mary, *Mary Thomas's Knitting Book,* London: Hodder and Stoughton, Ltd., 1938; New York: Joan Toggit, Ltd.

Walker, Barbara, *A Treasury of Knitting Patterns,* New York: Charles Scribner's Sons, 1968.

Walker, Barbara, *A Second Treasury of Knitting Patterns,* New York: Charles Scribner's Sons, 1970.

CROCHET

Blackwell, Liz, *A Treasury of Crochet Patterns,* New York: Charles Scribner's Sons, 1971.

Edson, Nicki Hitz and Stimmel, Arlene, *Creative Crochet,* New York: Watson-Guptill, 1973.

Hairpin Lace Instruction Leaflet #7685, Chicago: The Boye Needle Company.

MacKenzie, Clinton D., *New Design in Crochet,* New York: Van Nostrand Reinhold Company, 1972.

Taylor, Gertrude, *America's Crochet Book,* New York: Charles Scribner's Sons, 1972.

The Golden Hands Complete Book of Knitting and Crochet, New York: Random House, 1973.

BRAIDING

Cooke, Viva and Sampley, Julia, *Palmetto Braiding and Weaving,* Peoria: The Manual Arts Press, 1947.

D'Harcourt, Raoul, et al, eds., *The Textiles of Ancient Peru and Their Techniques.* Trans. Brown, Sadie; reprint, Seattle: University of Washington Press, 1962.

Grant, Bruce, *Encyclopedia of Rawhide and Leather Braiding,* Cambridge, Maryland: Cornell Maritime Press, Inc., 1972.

Graumont, Raoul and Hensel, John, *Encyclopedia of Knots and Fancy Rope Work,* Cambridge, Maryland: Cornell Maritime Press, 1958.

Shaw, George Russell, *Knots Useful and Ornamental,* New York: Bonanza Books, 1933.

Turner, Alta R., *Finger Weaving: Indian Braiding,* New York: Sterling, 1973.

BOBBIN LACE

DeDillmont, Thérèse, *Encyclopedia of Needlework,* Mulhouse, France: DMC Library, revised edition; New York: Joan Toggett, Ltd.

Nyrop-Larsen, Johanne, *Knipling Efter Tegning,* (Lace-Making by Diagram), Copenhagen: Jul. Gjellerups Forlag, 1955.

SPRANG

Hald, Margrethe, *Olddanske Tekstiler (Ancient Danish Textiles),* Copenhagen: Gyldendal, 1950.

Marein, Shirley, *Off the Loom,* New York: Viking, A Studio Book, 1972.

Skowronski, Hella and Reddy, Mary, *Sprang,* New York: Van Nostrand Reinhold, 1974.

BASKETRY

Cane and Basket Supply Company, 1283 South Cochran Avenue, Los Angeles, California 90019: Instruction booklets on all types of caning and rushing available at minimal cost.

Gallinger, Osma Couch, *Handweaving with Reeds and Fibers,* New York and London: Pitman, 1948.

Harvey, Virginia I., *The Techniques of Basketry,* New York: Van Nostrand Reinhold, 1974.

Krøncke, Grete, *Weaving with Cane and Reed,* New York: Van Nostrand Reinhold, 1967.

Maynard, Barbara, *Modern Basketry from the Start,* New York: Charles Scribner's Sons, 1973.

Perkins, H. H., *Instructions in Methods of Seat Weaving,* New Haven: The H. H. Perkins Company.

Rossbach, Ed, *Baskets as Textile Art,* New York: Van Nostrand Reinhold, 1974.

Sunset Books, eds., *Furniture Upholstery and Repair,* Menlo Park, California: Lane Books, 1970; "Caning and Rushing," pp. 74–76.

Index